AF361868

Health-Promoting Effects of Traditional Foods

Health-Promoting Effects of Traditional Foods

Editor

Marcello Iriti

MDPI • Basel • Beijing • Wuhan • Barcelona • Belgrade • Manchester • Tokyo • Cluj • Tianjin

Editor
Marcello Iriti
Department of Agricultural and
Environmental Sciences,
Milan State University
Italy

Editorial Office
MDPI
St. Alban-Anlage 66
4052 Basel, Switzerland

This is a reprint of articles from the Special Issue published online in the open access journal *Foods* (ISSN 2304-8158) (available at: https://www.mdpi.com/journal/foods/special_issues/health_effects_traditional_foods).

For citation purposes, cite each article independently as indicated on the article page online and as indicated below:

LastName, A.A.; LastName, B.B.; LastName, C.C. Article Title. *Journal Name* **Year**, *Article Number*, Page Range.

ISBN 978-3-03943-312-4 (Hbk)
ISBN 978-3-03943-313-1 (PDF)

Contents

About the Editor

Marcello Iriti is a Professor of Plant Biology and Pathology at the Department of Agricultural and Environmental Sciences, Faculty of Agricultural and Food Sciences, Milan State University. He has been studying bioactive phytochemicals relevant for human nutrition and health, including melatonin, polyphenols, carotenoids, sterols and essential oils, focusing on their functional role in planta, as well as on their in vitro/in vivo and in human biological activities. Furthermore, he has been investigating the effects of elicitors on plant secondary metabolism, as an approach to obtain phytochemical-enriched plant foods and medicinal herbs. He is the author of more than 200 publications (H-index: 41), as well as a member of the Asian Council of Science Editors, the Society of African Journal Editors, and he is a Founding Member of the Italian Society of Environmental Medicine. Furthermore, he is a member of the working group 'Pharmacognosy and Phytotherapy' of the Italian Pharmacological Society. His main patent is: 'Compositions Comprising Rutin Useful for the Treatment of Tumors Resistant to Chemotherapy' (WO2015036875A1; US20160213698; US9757405B2; EP3043821).

Editorial

Healthy Diets and Modifiable Risk Factors for Non-Communicable Diseases—The European Perspective

Marcello Iriti [1],*, Elena Maria Varoni [2],†ca and Sara Vitalini [1],†

[1] Department of Agricultural and Environmental Sciences, Milan State University, Via G. Celoria 2, 20133 Milan, Italy; sara.vitalini@unimi.it

[2] Department of Biomedical, Surgical and Dental Sciences, Milan State University, Via Beldiletto 1/3, 20122 Milan, Italy; elena.varoni@unimi.it

* Correspondence: marcello.iriti@unimi.it

† The two authors contribute to work equally.

Received: 27 June 2020; Accepted: 14 July 2020; Published: 16 July 2020

Abstract: Non-communicable diseases pose a serious threat to Western countries, in particular to European populations. In this context, healthy diets, such as the Mediterranean diet and the New Nordic diet developed in 2004, in addition to other healthy lifestyle choices (i.e., regular and low to moderate intensity levels of physical activity) can contribute to reduce the risk factors associated with cardiovascular disease and type 2 diabetes (majorly preventable, diet-related, non-communicable diseases), including being overweight, obesity, hypertension, hyperglycemia and hypercholesterolemia. The Mediterranean diet and the Nordic diet share common traits: they are rich in nutrient-dense foods (mostly plant-derived foods) and low in energy-dense foods (mainly of animal origin). However, more studies are needed to ascertain the long-term effects of adherence to both dietary styles with regards to disease prevalence and incidence, especially for the New Nordic Diet.

Keywords: Mediterranean diet; Nordic diet; overweight; obesity; cardiovascular disease; functional foods; nutraceuticals; bioactive phytochemicals

---

In the few last decades, the benefits of dietary styles rich in nutrient-dense foods have been emphasized in terms of longevity, healthy ageing and morbidity. Indeed, diets including plenty of plant foods have been associated with a reduced risk and incidence of chronic degenerative diseases such as cardiovascular disease, type 2 diabetes, metabolic syndrome, neurodegenerative disorders and some cancers. In general, healthy dietary habits include a low consumption of refined sugars, salt, saturated and trans fats, as well as high intake of fruit, vegetables (including legumes, whole grain cereals and nuts), low-fat dairy products and healthy lipids (from plant oils and seafood).

According to the World Health Organization (WHO), non-communicable diseases, including cardiovascular disease, cancers, respiratory diseases and diabetes, are responsible for almost 70% of all deaths worldwide. The rapid rise of non-communicable diseases has been driven by a number of (modifiable) behavioral risk factors, such as unhealthy diets, physical inactivity, exposure to tobacco smoke and the harmful use of alcoholic beverages, in addition to environmental (air pollutants), occupational (carcinogens, particulates, gases, fumes) and metabolic (overweight/obesity, hypertension, hyperglycemia, hypercholesterolemia) risk factors [1]. Of the six WHO regions, the WHO European Region is the most severely affected by non-communicable diseases (Table 1). Therefore, The European Food and Nutrition Action Plan 2015–2020 aimed at significantly reducing the burden of preventable diet-related non-communicable diseases, obesity and all other forms of malnutrition prevalent in the region, with emphasis on the decrease in the prevalence of obesity and diabetes, as well as overweight children under five years old [1].

Table 1. Burden of non-communicable diseases, overweight and obesity in the WHO European Region: factsheet.

- Cardiovascular disease, diabetes, cancer and respiratory diseases (the four major NCDs) together account for 77% of the burden of disease and almost 86% of premature mortality
- In 46 countries (accounting for 87% of the Region), more than 50% of adults (aged ≥ 20 years, both sexes) are overweight or obese, and in several countries the rate is close to 70% of the adult population
- Overweight and obesity are estimated to result in the death of approximately 320,000 men and women in 20 western European countries every year
- Rates of overweight and obesity in some parts of eastern Europe have risen more than threefold since 1980
- Overweight and obesity are also highly prevalent among children and adolescents, particularly in southern European countries
- The prevalence of overweight and obesity was 11–33% for children aged 11 years, 12–27% for children aged 13 years and 10–23% for those aged 15 years

Adapted from [1].

A number of studies have established the health-promoting effects of two European diets: the Mediterranean diet and the Nordic diet, particularly against cardiovascular disease and type 2 diabetes [2]. As substantiated by many observational studies, the Mediterranean diet, low in energy-dense foods, can be considered the archetype of a health-promoting lifestyle by virtue of the phytochemical diversity of its food components (Table 2).

Table 2. Dietary pattern and lifestyles of traditional Mediterranean diet.

Dietary Habits
1. High consumption of minimally-processed, local and seasonal plant food (whole-grain cereals, fresh fruit, cooked and raw vegetables, nuts)
2. Daily fat intake ranging from 25% to 35% of energy (with saturated fat ranging from ≤7% to 8% of energy)
3. Daily intake of low to moderate amounts of dairy products (mainly low-fat cheeses and yogurt)
4. Twice-weekly consumption of low to moderate amounts of fish and poultry; up to seven eggs per week
5. Fresh fruit as the typical dessert, with sweets containing sugars or honey consumed only a few times per week
6. Consumption of red meat only a few times per month
7. Regular, low to moderate consumption of wine at main meals; approximately 1-2 glasses per day for men and 1 glass for women (optional)
8. Herbs and spices to season food rather than salt or fat
Lifestyles
1. Regular daily physical activity
2. Enjoy meals with others (family and friends)

Adapted from [3].

The traditional Mediterranean diet originated in the olive- and grapevine-growing areas of the Mediterranean region and has a strong cultural association with these areas. It is characterized by a high intake of plant-based foods (cereals, fruit, vegetables, legumes and nuts) and olive oil; a moderate intake of fish and poultry; a low to moderate intake of red wine; and a low intake of dairy products (principally yogurt and cheese), red meat, processed meats and sweets (to which fresh fruit is often substituted). Social and cultural factors closely associated with the traditional Mediterranean diet, including shared eating practices, post-meal siestas (afternoon naps) and lengthy meal times, are also thought to contribute to the attributed positive health effects recorded in the Mediterranean region. However, the Mediterranean diet varies by country and region, despite the common traits, due to the climatic, cultural and religious differences among southern European, northern African and eastern Mediterranean populations [4].

The New Nordic Diet was developed in 2004 by scientists, nutritionists and chefs to address the growing overweight population and obesity rates, as well as the unsustainable farming systems in the Nordic countries (Table 3).

Table 3. Guidelines of the New Nordic Diet.

1. Eat more fruit and vegetables every day
2. Eat more whole grain products
3. Eat more food from the sea and lakes
4. Eat higher-quality meat, and less of it
5. Eat more food from wild landscapes
6. Eat organic products whenever possible
7. Avoid food additives
8. Eat more meals based on seasonal products
9. Eat more home-cooked food
10. Produce less waste

Adapted from [5].

This dietary style shares many characteristics with the Mediterranean diet, but comprises traditional foods from Denmark, Finland, Iceland, Norway and Sweden (please visit the Baltic Sea Diet Pyramid created by the Finnish Heart Association, the Finnish Diabetes Association and the University of Eastern Finland at www.helsinkitime.fi). Staple components of the New Nordic Diet include whole grain cereals (barley, oats and rye), vegetables (cabbage, tubers and root vegetables), legumes (mainly beans and peas), berries and fruit, nuts and seeds, and fish (herring, mackerel and salmon). A notable point of difference is the use of rapeseed (canola) oil instead of olive oil, rich in α-linolenic acid (a type of omega-3 polyunsaturated fatty acid). The Nordic diet is also characterized by a moderate consumption of dairy products and eggs, as well as a low intake of processed foods, sweets (including added sugars and sweetened beverages) and red meat. Not least, the Nordic diet is predominantly plant-based and locally sourced, thus ensuring a more environmentally friendly production with reduced waste when consumed within the Nordic region [2].

The health benefits of the Nordic diet have also been investigated—though to a lesser extent than those of the Mediterranean diet—and associated with improvements in risk factors for both cardiovascular disease and type 2 diabetes. In hypercholesterolemic individuals, the Nordic diet improved their blood lipid profile and insulin sensitivity, in addition to reducing blood pressure [6]. In subjects with metabolic syndrome, the Nordic diet ameliorated the blood lipid profile with beneficial effects on low-grade inflammations [7], besides decreasing the ambulatory blood pressure [8]. These findings, based on randomized clinical trials, were partially confirmed in population-based studies and the association between the adherence to the Nordic diet and cardiometabolic risk factors is still equivocal [9]. Adherence to the Nordic diet was also inversely associated with the risk of type 2 diabetes [10] and also induced weight loss in centrally obese men and women [11]. However, despite the fact that studies have shown that Nordic diet has beneficial effects on the risk factors for diabetes, such as obesity and low-grade inflammation, evidence on the long-term impact of adherence to the Nordic diet on diabetes prevalence and incidence requires larger prospective studies [12].

In conclusion, we have to take into account that the adherence to Mediterranean and Nordic diet may not always be high in the southern and northern European populations, respectively. In other words, the prevalence and mortality rate of cardiovascular disease and diabetes can be high in some of these countries (Figures 1 and 2). Therefore, to assess the adherence to both diets is pivotal in order to evaluate their predictive ability on specific risk factors and biomarkers, by powerful tools such as the Mediterranean diet score [13] and the Baltic Sea diet score [14].

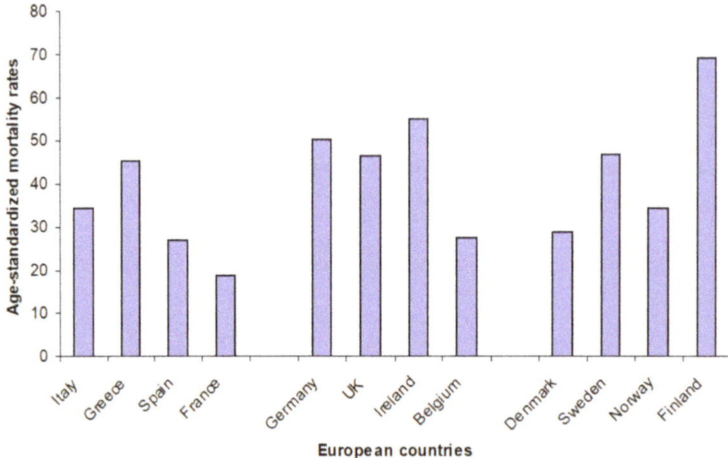

Figure 1. Age-standardized mortality rates per 100,000 population in 2014—ischemic heart diseases (both sexes) in selected European countries (from: WHO Mortality Database, https://apps.who.int/healthinfo/statistics/mortality/whodpms/).

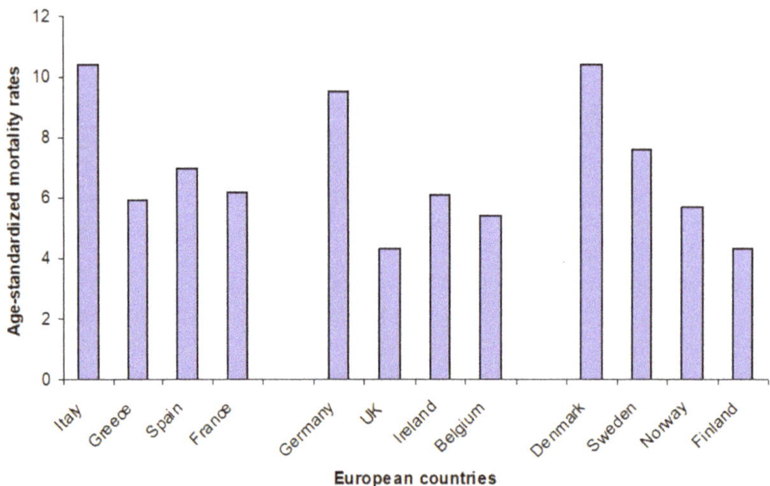

Figure 2. Age-standardized mortality rates per 100,000 population—diabetes mellitus (both sexes) in selected European countries (from: WHO Mortality Database, https://apps.who.int/healthinfo/statistics/mortality/whodpms/).

Author Contributions: Conceptualization, M.I.; data curation, M.I., S.V. and E.M.V.; writing—original draft preparation, M.I.; writing—review and editing, M.I., S.V. and E.M.V. All authors have read and agreed to the published version of the manuscript.

Funding: This research received no external funding.

Conflicts of Interest: The authors declare no conflict of interest.

References

1. World Health Organization—Regional Office for Europe. European Food and Nutrition Action Plan 2015–2020. Available online: https://www.euro.who.int/__data/assets/pdf_file/0003/294474/European-Food-Nutrition-Action-Plan-20152020-en.pdf (accessed on 20 June 2020).
2. Renzella, J.; Townsend, N.; Jewell, J.; Breda, J.; Roberts, N.; Rayner, M.; Wickramasinghe, K. *What National and Subnational Interventions and Policies Based on Mediterranean and Nordic Diets are Recommended or Implemented in the WHO European Region, and is There Evidence of Effectiveness in Reducing Noncommunicable Diseases? Health Evidence Network (HEN) Synthesis Report 58*; WHO Regional Office for Europe: Copenhagen, Denmark, 2018.
3. Shen, J.; Wilmot, K.A.; Ghasemzadeh, N.; Molloy, D.L.; Burkman, G.; Mekonnen, G.; Gongora, M.C.; Quyyumi, A.A.; Sperling, L.S.; Gongora, C.M. Mediterranean Dietary Patterns and Cardiovascular Health. *Annu. Rev. Nutr.* **2015**, *35*, 425–449. [CrossRef]
4. Iriti, M.; Vitallini, S.; Vitalini, S. Health-Promoting Effects of Traditional Mediterranean Diets—A Review. *Pol. J. Food Nutr. Sci.* **2012**, *62*, 71–76. [CrossRef]
5. Mithril, C.; Dragsted, L.O.; Meyer, C.; Blauert, E.; Holt, M.K.; Astrup, A. Guidelines for the New Nordic Diet. *Public Health Nutr.* **2012**, *15*, 1941–1947. [CrossRef] [PubMed]
6. Adamsson, V.; Reumark, A.; Fredriksson, I.-B.; Hammarström, E.; Vessby, B.; Johansson, G.; Risérus, U. Effects of a healthy Nordic diet on cardiovascular risk factors in hypercholesterolaemic subjects: A randomized controlled trial (NORDIET). *J. Intern. Med.* **2010**, *269*, 150–159. [CrossRef] [PubMed]
7. Uusitupa, M.; Hermansen, K.; Savolainen, M.J.; Schwab, U.; Kolehmainen, M.; Brader, L.; Mortensen, L.S.; Cloetens, L.; Johansson-Persson, A.; Onning, G.; et al. Effects of an isocaloric healthy Nordic diet on insulin sensitivity, lipid profile and inflammation markers in metabolic syndrome —A randomized study (SYSDIET). *J. Intern. Med.* **2013**, *274*, 52–66. [CrossRef] [PubMed]
8. Brader, L.; Uusitupa, M.; Dragsted, L.O.; Hermansen, K. Effects of an isocaloric healthy Nordic diet on ambulatory blood pressure in metabolic syndrome: A randomized SYSDIET sub-study. *Eur. J. Clin. Nutr.* **2013**, *68*, 57–63. [CrossRef] [PubMed]
9. Kanerva, N.; Kaartinen, N.E.; Rissanen, H.; Knekt, P.; Eriksson, J.G.; Sääksjärvi, K.; Sundvall, J.; Männistö, S. Associations of the Baltic Sea diet with cardiometabolic risk factors—A meta-analysis of three Finnish studies. *Br. J. Nutr.* **2014**, *112*, 616–626. [CrossRef] [PubMed]
10. Lacoppidan, S.A.; Kyrø, C.; Loft, S.; Helnæs, A.; Christensen, J.; Hansen, C.P.; Dahm, C.C.; Overvad, K.; Tjonneland, A.; Olsen, A. Adherence to a Healthy Nordic Food Index Is Associated with a Lower Risk of Type-2 Diabetes—The Danish Diet, Cancer and Health Cohort Study. *Nutrients* **2015**, *7*, 8633–8644. [CrossRef] [PubMed]
11. Poulsen, S.K.; Due, A.; Jordy, A.B.; Kiens, B.; Stark, K.D.; Stender, S.; Holst, C.; Astrup, A.; Larsen, T.M. Health effect of the New Nordic Diet in adults with increased waist circumference: A 6-mo randomized controlled trial. *Am. J. Clin. Nutr.* **2013**, *99*, 35–45. [CrossRef] [PubMed]
12. Kanerva, N.; Rissanen, H.; Knekt, P.; Havulinna, A.; Eriksson, J.G.; Männistö, S. The healthy Nordic diet and incidence of Type 2 Diabetes—10-year follow-up. *Diabetes Res. Clin. Pract.* **2014**, *106*, e34–e37. [CrossRef] [PubMed]
13. Panagiotakos, D.B.; Pitsavos, C.; Stefanadis, C. Dietary patterns: A Mediterranean diet score and its relation to clinical and biological markers of cardiovascular disease risk. *Nutr. Metab. Cardiovasc. Dis.* **2006**, *16*, 559–568. [CrossRef] [PubMed]
14. Kanerva, N.; Kaartinen, N.E.; Schwab, U.; Lahti-Koski, M.; Männistö, S. The Baltic Sea Diet Score: A tool for assessing healthy eating in Nordic countries. *Public Health Nutr.* **2013**, *17*, 1697–1705. [CrossRef] [PubMed]

 *foods*

Review

Patagonian Berries: Healthy Potential and the Path to Becoming Functional Foods

Lida Fuentes [1,*]**, Carlos R. Figueroa** [2]**, Monika Valdenegro** [3] **and Raúl Vinet** [1,4]

[1]   Centro Regional de Estudios en Alimentos Saludables (CREAS), CONICYT-Regional GORE Valparaíso Proyecto R17A10001, Avenida Universidad 330, Placilla, Curauma, Valparaíso 2340000, Chile
[2]   Institute of Biological Sciences, Campus Talca, Universidad de Talca, Talca 3465548, Chile
[3]   Escuela de Agronomía, Facultad de Ciencias Agronómicas y de los Alimentos, Pontificia Universidad Católica de Valparaíso, Calle San Francisco s/n, Casilla 4-D, Quillota 2260000, Chile
[4]   Laboratorio de Farmacología, Centro de Micro Bioinnovación (CMBi), Facultad de Farmacia, Universidad de Valparaíso, Gran Bretaña 1093, Valparaíso 2360102, Chile
*   Correspondence: lfuentes@creas.cl; Tel.: +56-32-237-2868

Received: 30 June 2019; Accepted: 22 July 2019; Published: 26 July 2019

Abstract: In recent years, there has been an increasing interest in studying food and its derived ingredients that can provide beneficial effects for human health. These studies are helping to understand the bases of the ancestral use of several natural products, including native fruits as functional foods. As a result, the polyphenol profile and the antioxidant capacity of the extracts obtained from different Patagonian native berries have been described. This review aims to provide valuable information regarding fruit quality, its particular compound profile, and the feasibility of producing functional foods for human consumption to prevent disorders such as metabolic syndrome and cardiovascular diseases. We also discuss attempts concerning the domestication of these species and generating knowledge that strengthens their potential as traditional fruits in the food market and as a natural heritage for future generations. Finally, additional efforts are still necessary to fully understand the potential beneficial effects of the consumption of these berries on human health, the application of suitable technology for postharvest improvement, and the generation of successfully processed foods derived from Patagonian berries.

Keywords: maqui; murta; calafate; arrayán; Chilean strawberry; berries; functional foods

---

1. Introduction

When we imagine a place like Patagonia, it is impossible not to evoke images of extraordinary beauty like southern ice fields. However, a walk through this place also allows us to contemplate ancestral traditions that include the use of many native species. This southernmost region of the South American continent extends from 37° S to Cape Horn, at 56° S, whose geography is characterized by the Andes range, which is both the continental watershed and the international limit between Argentina and Chile. It includes the Pacific and Atlantic coasts and lowlands, the southern archipelagos and tablelands, and the valleys and high plains extending between the Andes and the Atlantic Ocean [1].

The Andean temperate forests of Patagonia have a great diversity of plants with medicinal properties [2,3]. The use of medicinal and edible native plants is a long-standing tradition in the Mapuche communities of Southern Argentina and Chile [4–6]. An ethnobotanical survey conducted in rural villages of San Martin de Los Andes, Argentina, showed the use and knowledge of about 40 and 47 native plants, respectively [5]. Unfortunately, this ancient knowledge tends to disappear in the younger generations [5]. Moreover, the effects of human activity (e.g., an increase in dwelling number) and the invasion of alien plants can reduce the availability of forest-associated gathering sites. Therefore, the use of food derived from non-cultivated plants as part of the diet could be a tradition susceptible to disappearing [7–9] and the cultural, social, and economic aspects must be evaluated comprehensively if these traditions are to be maintained for future generations [8,9].

In recent years, the interest in food or ingredients that provide beneficial effects for human health has increased. As a result, many native fruits from different continents have been studied as a source of functional foods [9–16]. In Chilean Patagonia, edible fruits come from woody or shrub forest species belonging to the Elaecarpaceae, Berberidaceae, and mainly Myrtaceae families [15,16], and creeping plants belong to Rosaceae family. These species present fruits rich in antioxidant and functional compounds, such as *Aristotelia chilensis* (maqui), *Berberis microphylla* (calafate), *Ugni molinae* (murta), *Luma apiculata* (arrayán), and *Fragaria chiloensis* (Chilean strawberry), among others [15–23] (Table 1). In Chile, these native species are mainly distributed from the Coquimbo to Magallanes regions (Latitude 31° to 55°), with Chilean Patagonia being the common region for all fruits analyzed in the present review (Table 1).

Most of the traditional uses of these fruits include consumption as fresh and dried fruits or being used to make tea, jam, cakes, juice, alcoholic beverages, and textile tinctures. Moreover, they have tremendous functional potential due to their high antioxidant values, particularly flavonol and anthocyanin contents and promissory bioassay results as anti-inflammatory, antidiabetic, and hypolipidemic agents [11,15,16,20–27]. Recently, the morphological characterization, geographical distribution, and ethnobotany of many of these species have been described in detail by Ulloa-Inostroza et al. (2017) [15] and Schmeda-Hirschmann et al. (2019) [16]. In this review, we focus on five fruit species growing in Patagonia with high potential as functional food (i.e., maqui, murta, calafate, arrayán, and Chilean strawberry, see Table 1); giving a little background on the fruit quality; and discussing the recent research data available—regarding the particular compound profile, their processing, and clinical assays—and the aspects to consider the commercial prospection of these Patagonian berries.

2. Quality Aspects and Bioactive Compounds of Patagonian Berries

2.1. Fruit Quality

According to Barrett et al. (2010) [28], in reference to fruits, the characteristics that impart a distinctive quality may be described by four different attributes: color and appearance, flavor (taste and aroma), texture, and nutritional value. All these aspects are determined through the complex biological process of fruit development and ripening [29,30].

We next summarize the main quality aspects of Patagonian berries, such as color and appearance, flavor, and texture. Nutritional and functional value-related antecedents of berries will be addressed in the next sections.

2.1.1. Color and Appearance

The precise definition of the developmental and ripening stages is necessary to determine the physicochemical and physiological parameters that contribute to the different quality attributes of fruit at harvest. A representative fruit at the ripe stage for each species analyzed in this review is shown in Figure 1. In Chilean strawberry, four developmental fruit stages have been described (i.e., small green, C1; large green, C2; turning, C3; and ripe fruit, C4) [31]. The ripe fruit stage (at harvest) has shown a pink receptacle and red achenes that, in comparison with the ripe stage of *Fragaria* x *ananassa* ('Aromas') fruit, can be 200-fold less red (comparison of the a* color parameter) [32,33]. Regarding the fruit weight of *F. chiloensis* fruit, it is nearly half of that present in modern commercial strawberry varieties (such *F.* x *ananassa* 'Chandler') [31]. In maqui berry, five different maturity stages have been described, starting from 21 days after fruit set in central Chile and named as green (I and II), light red, purple, and dark purple stages [34]. The berry weight per 100 fruits ranges from 10 g (green I stage) to 21 g (dark purple stage), with the highest increase in weight between the green II and light red stages.

Table 1. Main features of Patagonian fruits analyzed in the present review. Scientific and common names, botanic family, geographic distribution, traditional products and uses, and functional products generated in the last years.

Species	Common Name	Family	Geographic Distribution [16,35]	Traditional Products and Uses	Functional Products
Aristotelia chilensis (Mol.) Stuntz.	Maqui	Elaeocarpaceae	Chile: from the Coquimbo to Aysén regions, including Juan Fernández Island (Latitude 31°–40°). Argentina: from Jujuy to Chubut provinces.	Fresh and dried fruit, use to make textile pigment, cake, jam, juice, alcoholic beverages [36,37]	Freeze-dried maqui (powder and capsules), honey mix, functional drinks, drugs [24–26,38–41]
Ugni molinae Turcz.	Murta	Myrtaceae	Chile: From the O'Higgins to Aysén regions, including Juan Fernández Island (Lat. 34°–40°). Argentina: Neuquén, Rio Negro, and Chubut provinces.	Fresh and dried fruit, textile pigment, bakery, jam, alcoholic beverages [37]	Freeze-dried murta (powder and capsules), honey mix [41,42]
Berberis microphylla G. Forst.	Calafate	Berberidaceae	Chile: From the Metropolitan to Magallanes regions (Lat. 33°–55°). Argentina: From Neuquén to Tierra del Fuego provinces.	Fresh fruit, used to make jam, juice, beer [36,37]	Natural colorants [37]
Luma apiculata (DC.) Burret.	Arrayán	Myrtaceae	Chile: From the Coquimbo to Aysén regions (Lat. 31°–40°). Argentina: From Neuquén to Chubut provinces.	Fresh fruit, textile pigment, bakery, jam, aromatic wine [22,23]	N.D.
Fragaria chiloensis (L.) Mill.	Chilean strawberry	Rosaceae	Chile: From the O'Higgins to Magallanes regions (Lat. 34°–55°). Argentina: Neuquén and Rio Negro provinces.	Fresh fruit, used to make alcoholic beverages, cake [36,43]	N.D.

Geographic distribution according to Rodriguez et al., 2018 [35] and Schmeda et al., 2019 [16]. N.D.: not described.

Figure 1. Patagonian berries with healthy potential as a functional food on the basis of recent research data available. (**A**) *Aristotelia chilensis* (Mol.) Stuntz (maqui)*; (**B**) *Ugni molinae* Turcz. (murta)*; (**C**) *Berberis microphylla* G. Forst. (calafate)*; (**D**) *Luma apiculata* (DC.) Burret (arrayan)**; (**E**) *Fragaria chiloensis* (L.) Mill. (Chilean strawberry)**. Photography credit to M. Teresa Eyzaguirre-Philippi (*) and Carlos R. Figueroa (**).

The shape of murta and arrayán fruit was reported as globular, with a major equatorial diameter [21,44]. As far as we know, the only report for arrayán fruit development was made by Fuentes et al. (2016) [21]. In that work, the authors classified the fruit development into four stages, mainly by fruit shape and skin color: small and thin green (La1), rounded turning (La2), rounded purple (La3), and black ripe (La4) berries, with a decrease in lightness (L*), b*, and chroma values of fruit skin from approximately 48 to 23 from La1 to La4. As expected, a constant increase in the fresh and dry weight was observed during fruit development, although a higher increment was noted between the La3 and La4 stage [21].

For calafate berry, Arena and Curvetto (2008) [45] described a typical double sigmoid curve for fruit growth, with a constant increase in both the fresh weight and the diameter from 14 to 84 days after full flowering, reaching a maximum of 420 mg and 9.6 mm, respectively.

2.1.2. Flavor

The soluble solids content (SSC) and titratable acidity (TA) are proper predictor parameters for the ripening in several fleshy fruits and are the main determinants for fruit flavor [46,47]. Generally, a reduction of TA and a concomitant increase in SSC are observed during fleshy fruit ripening [48]. Calafate, Chilean strawberry, and arrayán fruits presented this pattern from green to ripe stages [21,49–51], but no extensive information has existed in maqui and murta berries until now. The SSC/TA ratio in the arrayán berry increases significantly in the final stages of development [21]. In maqui berry, soluble solids increase during ripening and in dark purple stages, range from 18.8 to 19.9° Brix [34], whereas in the ripe stage of arrayán (black ripe stage), a range between 11.5 to 12.5° Brix was observed [21]. In murta, 22–25% SSC, 4–8 g/L of organic acid (tartaric), and pH 4.7–5.2 were reported in ripe fruit [44]. For the calafate berry, the entire fruit growth period reaches up to 126 days from the full flower, where the fruit presents the highest SSC (25° Brix) and the lowest TA (2.19% to 2.6%) values [45,49]. In this sense, a 4.5-fold increase in the SSC/TA ratio was observed from 56 to 126 days from a full flower in the calafate berry [45]. In calafate, citric and malic acid contents increased during the first stages of fruiting and then decreased toward the end of ripening, although the citric acid content stayed constant from the onset of ripening. Oxalic and tartaric acid contents were maximal between 42 and 70 days from a full flower and then decreased toward the end of the fruiting period [50].

Relatively little information is available regarding the aroma profiles of maqui, calafate, and arrayán berries. In contrast, more detailed information could be found for Chilean strawberry and murta [51,52]. Chilean strawberry fruit is characterized by its great aroma and flavor [51,53]. In this sense, González et al. (2009) [51] identified mainly esters, and secondary alcohols and ketones, with esters and alcohols being up to 73% and 25% of the total volatiles at the ripe stage, respectively. Some esters were reported for the first time in Chilean native strawberry, without references in the commercial strawberry [51], suggesting that the native species has a particular aroma profile. In murta, aroma evolution during storage showed 24 volatile compounds identified, and the concentration of these compounds ranged from 1.2 to 250.5 µg kg^{-1} fresh weight. Methyl 2-methyl butanoate, ethyl butanoate, ethyl hexanoate, ethyl benzoate, ethyl 2-methyl butanoate, methyl hexanoate, and methyl benzoate were the major components, while the most potent compounds in the murtilla fruit aroma were ethyl hexanoate and 4-methoxy-2,5-dimethyl-furan-3-one [52].

2.1.3. Texture

Fruit firmness is one of the leading quality attributes of texture and has an essential commercial impact for both exporters and consumers [54]. In this sense, firmness should be a key goal of breeding in Patagonian soft berries. Strawberry is one of the softest fruits, and loss of firmness is well-documented as being related to cell wall disassembly during ripening [55]. Most significant decreases in cell wall polymers associated with Chilean strawberry fruit ripening occur within the pectin fractions, especially in the covalently bound pectin fraction, which is highly correlated with firmness loss and an increase in the activity of specific cell wall-related enzymes, such as beta-galactosidase [55]. It was reported that the modified atmosphere packaging (MAP) of Chilean strawberry during 12 days of storage at 4 °C delayed the fruit dehydration and the firmness loss, that allow the preservation of quality parameters and anthocyanin compounds compare to fruit storage in the control conditions [56]. However, the application of MAP technology diminished the relative abundance of total volatile compounds [56]. In the arrayán berry, a significant reduction in fruit firmness was observed between rounded purple and black ripe stages [21], although this loss in firmness is slower than that observed during the fruit development of *F. chiloensis* [31]. The firmness reduction of *L. apiculata* fruit [21] showed similar values and trends reported for blueberry fruit [57]. A comparative study of postharvest in the two varieties of murta (i.e., South Pearl INIA and Red Pearl INIA) showed that Red Pearl INIA has a major shelf life during 35 days of storage at 0 °C [58]. The postharvest assay showed a storage capacity of South Pearl INIA during 20 days at 0 °C, while Red Pearl INIA showed major potential for post-harvesting [59]. During treatment of a controlled atmosphere (CA), Red Pearl INIA was stored without problems until 35 days, while South Pearl INIA showed storability until 25 days [59].

2.2. Antioxidant Capacity

In plants, phenolic compounds are produced as secondary metabolites exerting various protective roles and are generally involved in the defense against stress conditions [60–63]. The main phenolic compounds in these fruits can be divided into phenolic acids, and flavonoids such as flavonols, flavanols, and anthocyanins (Figure 2) [62,63]. These molecules are responsible for the major organoleptic characteristics of plant food, such as the visual appearance, flavor, bitterness, astringency, and aroma [64]. Many beneficial effects attributed to phenolic compounds [64–67] have given rise to a new interest in finding plant species with a high phenolic content and relevant biological activity. Studies on the phenolic compounds of the fruits of maqui, murta, calafate, arrayán, and Chilean strawberry highlight the high antioxidant activity they present [15–23] (Table 2). In the following section, we briefly summarize the available literature on the main phenolic compounds described for the Patagonian berries analyzed in this review (Table 2).

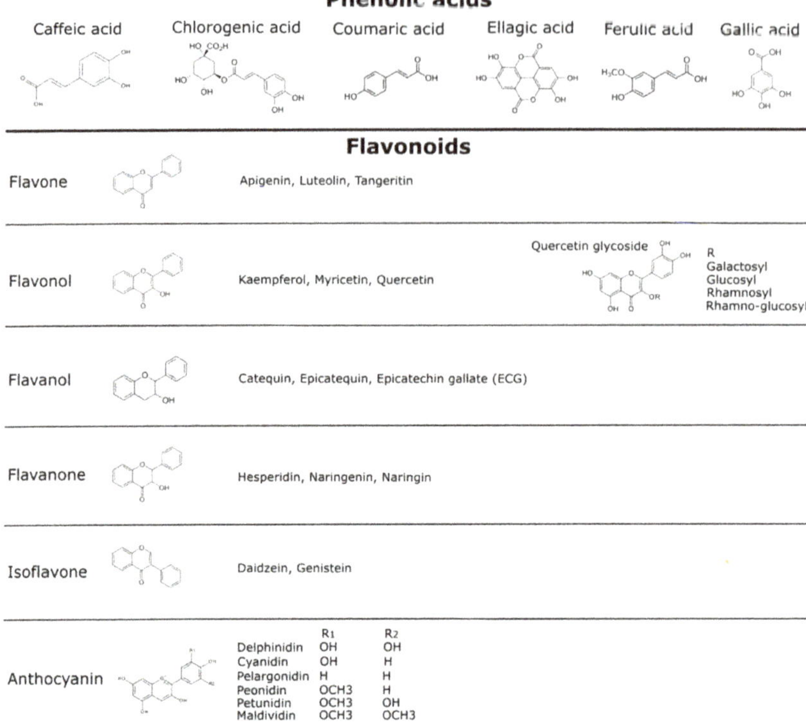

Figure 2. Polyphenols compounds described in vegetables and fruits. Different phenolic compounds have been reported in native Chilean berries, including phenolic acid, flavonoids such as quercetins—principally quercetin glycosides—and anthocyanins [15–23]. More details are presented in the text. Chemical structures credits [68].

Table 2. Antioxidant information of Patagonian berries.

Species Name	Average Antioxidant Capacity Determined by ORAC (μmol·100 g DW^{-1}) [a]	Average Range of Total Polyphenols Compounds Content (mg GAE g^{-1} DW^{-1}) [a]	Number of Non-Anthocyanin Polyphenol Compounds Reported	Principal Non-Anthocyanin Polyphenol Compounds	Number of Anthocyanin Compound Reported	Principal Anthocyanin Compounds
Maqui.	37,174 [11,69]	49.7 [70]	13 [15]	Quercetin, dimethoxy-quercetin, quercetin-3-rutinoside, quercetin-3-galactoside, myricetin and its derivatives (dimethoxy-quercetin) and ellagic acid [70]	8 [15]	3-glucosides, 3,5-diglucosides, 3-sambubiosides and 3-sambubioside-5-glucosides of cyanidin and delphinidin (delphinidin 3-sambubioside-5-glucoside) [20,71]
Murta	43,574 [11,69]	9.2 [19] 34.9 [69]	16 [15]	caffeic acid-3-glucoside, quercetin-3-glucoside, quercetin, gallic acid, quercetin-3-rutinoside, quercitrin, luteolin, luteolin-3-glucoside, kaempferol, kaempferol-3-glucoside, myricetin and p-coumaric acid [72]	11 [15]	delphinidin-3-, malvidin-3- and peonidin-3-arabinoside; peonidin-3- and malvidin-3-glucoside [20,72]
Calafate	72,425 [11,69]	33.9 [69] 65.5 [19]	36 [15]	quercetin-3-rutinoside, gallic- and chlorogenic acid, caffeic and the presence of coumaric- and ferulic acid, quercetin, myricetin, and kaempferol [19]	30 [15]	delphinidin-3-glucoside, delphinidin-3-rutinoside, delphinidin-3,5-dihexoside, cyanidin-3-glucoside, petunidin-3-glucoside, petunidin-3-rutinoside, petunidin-3,5-dihexoside, malvidin-3-glucoside and malvidin-3-rutinoside [19,20]
Arrayán	62,500 [21]	27.6 [19]	13 [15]	quercetin 3-rutinoside and their derivatives, tannins and their monomers [18,21]	8 [15]	peonidin-3-galactoside, petunidin-3-arabinoside, malvidin-3-arabinoside, peonidin-3-arabinoside delphinidin-3-arabinoside, cyanidin-3-glucoside, peonidin-3-glucoside and malvidin-3-glucoside [18,19,21]
Chilean strawberry	N.R.	N.R	16*20** [17]	ellagic acid and their pentoside- and rhamnoside derivatives. quercetin glucuronide, ellagitannin, quercetin pentoside, kaempferol glucuronide. Catechin *, quercetin pentosid *, and quercetin hexoside * procyanidin tetramers ** and ellagitannin ** [17]	4 [17]	cyanidin 3-O-glucoside, pelargonidin 3-O-glucoside, cyanidin-malonyl-glucoside and pelargonidin-malonyl- glucoside [17]

The table shows the available data concerning the antioxidant capacity determined by oxygen-radical absorbing capacity (ORAC) (μmol·100 gDW^{-1}), total polyphenols compounds content (mg GAE gDW^{-1}), and polyphenol compounds reported in these fruits. N.R.: not reported. (*) polyphenols compounds reported in *F. chiloensis* ssp. *chiloensis* f. chiloensis and reported in (**) *Fragaria chiloensis* ssp. *chiloensis* f. patagonica. More details are given in the text. [a] DW, dry weight; GAE, gallic acid equivalents.

Different methods have been used for determining the total antioxidants in different vegetables and fruit, including Patagonian berries. Currently, the oxygen-radical absorbing capacity (ORAC) is a method commonly used to compare the antioxidant capacity in different foods [11,73]. The ORAC values (as µmol per 100 g of dry weight, DW) of maqui (37,174), calafate (72,425), murta (43,574), and arrayan (62,500) berries were reported as being higher than in commercial berries such as raspberries, blueberries (*Vaccinium corymbosum* 'Bluegold') (27,412), and blackberries cultivated in Chile [11,21,69]. Similar trends were reported using different methods [20]. The Trolox equivalent (TE) antioxidant capacity (TEAC) showed that maqui (88.1) and calafate (74.5) had a higher antioxidant capacity (µmol TE per gram of fresh weight, FW) compared to murta (11.7) and blueberry (14.5) fruits [20]. The analysis by 2,2-diphenylpicrylhydrazyl (DPPH) methods showed that the antioxidant activity (mg of crude extract per liter) was higher in maqui (399.8) than in murta (82.9) [15]. The IC_{50} range of maqui extract (0.0012 and 0.0019 g L^{-1}) compared to the average value (0.03 g L^{-1}) of commercial berries cultivated in Chile, such as blueberry (*V. corymbosum*), strawberry (*F.* x *ananassa*), and raspberry (*Rubus idaeus*), indicates that a minor concentration of maqui extract is required to inhibit DPPH radicals [74,75]. The above information represents a fundamental background supporting the idea that the Patagonian berries have good potential as a functional food, by themselves or as food ingredients.

2.2.1. Phenolic Content and Composition

The phenolic compounds reported in native Chilean berries include caffeic acid, ferulic acid, gallic acid, myricetin, p-coumaric acid, and others [15–23]. Similar to what has been observed for the antioxidant capacity, high total polyphenols contents (TPC) were found for maqui and calafate [19,20]. The different reports of total phenolic analysis using the Folin–Ciocalteu method showed different rankings for Patagonian berries. The first studies showed a higher total phenol content (as µmol gallic acid equivalents (GAE) per gram of FW) for maqui (97 µmol GAE g^{-1} FW) and calafate (87 µmol GAE g^{-1} FW), followed by murta (32 µmol GAE g^{-1} FW) compared to blueberry (17 µmol GAE g^{-1} FW) [20]. Some reports showed similar values of the total polyphenols content (as mg GAE per gram of DW) for calafate (33.9 mg GAE g^{-1} DW), maqui (31.2 mg GAE g^{-1} DW) and murta (34.,9 mg GAE g^{-1} DW) [11], while other reports indicated significant differences between Patagonian berries, with higher values for calafate (65.5 mg GAE g^{-1} DW), followed by arrayán (27.6 mg GAE g^{-1} DW), and lower values for murta (9.2 mg GAE g^{-1} DW) [19].

Concerning the polyphenols composition of the Patagonian berries, maqui and calafate showed anthocyanin as the main component, while fruits of the Myrtaceae family (e.g., murta and arrayán) showed a higher content of flavonoid compounds [15,16,18–21,70–72]. Calafate fruit showed a comparable flavonoid content (0.16 µmol g^{-1} FW) to that obtained for maqui fruit (0.12 µmol g^{-1} FW) [20]. In calafate berry collected from different localities, the identification of flavonoids and phenolic acids showed a higher content of rutin, gallic-chlorogenic, and caffeic acid, and the presence of coumaric and ferulic acid, quercetin, myricetin, and kaempferol [20,76].

The multiple bioactive compounds of the maqui berry (i.e., phenolic antioxidants, alkaloids, flavonoids, and particularly anthocyanins) have contributed to knowledge of the functional potential of this berry in several countries [38,77–79]. An HPLC analysis of maqui berry extracts showed 10 compounds identified as flavonols and ellagic acid [70]. The non-anthocyanin compounds were mainly quercetin and its derivatives (with the highest concentration of dimethoxy-quercetin, followed by rutin (quercetin-3-rutinoside) and quercetin-3-galactoside), myricetin and its derivatives, and an important content of ellagic acid [70].

In arrayán, the polyphenol compounds identified mainly correspond to flavonols such as quercetin 3-rutinoside and its derivates, tannins and their monomers, and a minor number of anthocyanins [18,21]. In murta, caffeic acid-3-glucoside, quercetin-3-glucoside, and quercetin were reported as three major compounds in ethanolic extracts of fruit, and the others compounds were gallic acid, rutin, quercitrin, luteolin, luteolin-3-glucoside, kaempferol, kaempferol-3-glucoside, myricetin, and p-coumaric acid [72].

In the Chilean strawberry species, several compounds were identified, including an ellagic acid-based compound, catechin, and flavonol derivatives. The higher content of non-anthocyanins identified in *F. chiloensis* and *F. x ananassa* 'Chandler' were ellagic acid and their pentoside and rhamnoside derivatives and quercetin glucuronide [17]. On the other hand, ellagitannin, quercetin pentoside, and kaempferol glucuronide were only reported in *F. chiloensis* and some compounds—catechin, quercetin pentoside, and quercetin hexoside—were only reported in *Fragaria chiloensis* ssp. *chiloensis* f. chiloensis, and other compounds—procyanidin tetramers and ellagitannin—were only reported in *F. chiloensis* ssp. *chiloensis* f. patagonica [17].

2.2.2. Anthocyanins

Different studies suggest that the highest total anthocyanin content (TAC) can be found in calafate and maqui berries, especially those harvested in the Chilean Patagonia, followed by fruits of the Myrtaceae family species, i.e., arrayán and murta [15,20,80]. It was reported that the total anthocyanin concentrations were higher in calafate fruit extract (between 14 and 26 μmol g^{-1} FW) [20] and (between 23 and 36 μmol g^{-1} FW) [80], followed by maqui berries (between 16 and 20 μmol g^{-1} FW), whereas murta (0.2 μmol g^{-1} FW) showed lowest values than blueberry (2.0 μmol g^{-1} FW) [20].

Similar results were reported by Brito et al. (2014) [19], with a higher anthocyanin content (as mg cyanidin 3-O-glucoside g^{-1} DW) in calafate (51.6), followed by arrayán (15.2) and murta (6.9) berries. The anthocyanin composition of the maqui berry corresponds to 3-glucosides, 3,5-diglucosides, 3-sambubiosides, and 3-sambubioside-5-glucosides of delphinidin and cyanidin, and 34% of total anthocyanins correspond to delphinidin 3-sambubioside-5-glucoside, the major anthocyanin [71,81]. In calafate berry, the main anthocyanins described were delphinidin-3-glucoside, delphinidin-3-rutinoside, delphinidin-3,5-dihexoside, cyanidin-3-glucoside, petunidin-3-glucoside, petunidin-3-rutinoside, petunidin-3,5-dihexoside, malvidin-3-glucoside, and malvidin-3-rutinoside [20]. The above suggests that the antioxidant capacity observed in calafate berries is probably due to their anthocyanin diversity and, in maqui, is due to the particular presence of delphinidin 3-sambubioside-5-glucoside.

Nevertheless, the higher flavonoid content in the Myrtaceae family [11,20], anthocyanins such as peonidin-3-galactoside, petunidin-3-arabinoside, malvidin-3-arabinoside, and peonidin-3-arabinoside, were reported in both the methanol-HCl and methanol extracts of arrayán fruit [21]. The first three have been described in blueberry [82], and delphinidin-3-, malvidin-3-, and peonidin-3-arabinoside; peonidin-3- and malvidin-3-glucoside were described in murta and calafate berries [82]. Other anthocyanins, such as delphinidin-3-arabinoside, cyanidin-3-glucoside, peonidin-3-glucoside, malvidin-3-glucoside, and petunidin-3-arabinoside, were observed in a methanol-HCl extract of arrayán fruit by different authors [21,82].

The two major anthocyanins identified in both Chilean strawberry botanical forms were cyanidin 3-O-glucoside and pelargonidin 3-O-glucoside; these two compounds have generally been described in different *Fragaria* spp. [17]. On the other hand, cyanidin-malonyl-glucoside and pelargonidin-malonyl-glucoside were only reported in Chilean strawberries compared to commercial strawberry ('Chandler') [17].

3. Effects of Processing on Bioactive Compounds

Many native fruits are only available in determining seasons, so it is difficult to have these fresh fruits for consumption all year or away from collection sites. In general, anthocyanins are susceptible to degradation under environmental conditions, such as oxygen, heat, and changes in pH, among others [83]. The effectiveness, uniformity, and richness of these products are dependent upon the preservation of bioactive compounds throughout the value-added chain. Native berries exhibit high water activity and are highly perishable and susceptible to microbial deterioration, enzymatic reactions, and oxidation [39]. The effects of drying, the microencapsulation process, and juice preparation have been evaluated in maqui and murta berries. In addition, maqui and murta leaf extracts have been evaluated as ingredients to incorporate in food or coating. It was reported that the incorporation of murta leaves extracts in tuna-fish (*Thunnus tynnus*) gelatin-based edible films leads to transparent films with increased protection against UV light and antioxidant capacity [84]. The availability of new products based on maqui and murta as functional ingredients among other Patagonian berries goes hand in hand with the study of the preservation techniques of these fruits. In the following sections, we summarize the literature regarding the effect of processing, with an emphasis on functional maqui and murta products.

3.1. Drying Process

Advances in drying technology and standardization techniques in compound analysis allow for the possibility of using drying for the development of functional foods and nutraceuticals. It is essential to consider that the selection of the type of dryer or drying system used for a specific situation must be based upon the product's characteristics and drying behavior, as well as the end product required [85]. Solar drying (SoD) is the cheapest method for drying whole fruits and vegetables. However, the long drying time and the risk of contamination and spoilage due to exposure to an open environment are the main drawbacks associated with this method. Hot air dryers (HAD) are commonly used by the food industry as they provide relatively fast, uniform, and sanitary drying [86]. However, in most cases, it is possible to modify the richness of bioactive compounds of the raw material, as a function of the temperature/time combination applied in the process [87]. Freeze drying (FD) can produce high-quality products, but is comparatively more expensive; however, and despite this, FD use has been increasing in the industry of processing fruits [88].

Besides its potential role in the battle against certain illnesses and degenerative diseases, some native fruits like maqui or commercial varieties of species like blueberries, cranberries, and tomatoes, share a unique characteristic: a waxy outer skin. The waxy layer affects the flow of moisture from inside the fruit to its surface, which is a crucial process in drying. In the particular case of maqui fruits, the drying process is limited by an external waxy layer similar to that of grapes, which hinders mass water transfer and reduces the drying rate [89]. Technologies and methods applicable to the drying of small waxy skinned fruit could be suitable for obtaining foods and nutraceuticals from maqui fruits. In these cases, several chemical and physical pre-treatments were suggested by several authors to improve the drying rate of whole fruits with waxy skins, e.g., grapes, cherries, plums, apricots, and blueberries [90–95]. Pre-treatment methods employing chemical dipping, mechanical processes, and thermal treatments have been used to overcome the wax barrier in several applications [96–100].

Drying technologies are widely used in the industry as a strategy to protect functional molecules—anthocyanins—in value-added products, such as health food ingredients. Convective air drying technologies such as cabinets or trays, fluidized beds, spouted beds, and microwave/spouted beds (MWSB), and those using other technologies (spray-drying, freeze-drying, vacuum, microwave, and osmosis), are some alternatives for processing fruits and vegetables [101]. Between them, spray-drying (SD) is available in the pharmaceutical and food industries [83,102–104]. This method is the most used in the food industry because it is economical, rapid, and effective in protecting this compound [105]. For example, SD is widely used in the pharmaceutical and food industries to encapsulate anthocyanin compounds due to the short drying times (5–30 s) [83,102–104]. During the last decade, freeze-drying (FD) has become more widespread in the food industry [103]. The FD technique is based on the removal of water from a frozen product by sublimation and has been used as an alternative method to encapsulate anthocyanins [106]. An economically accessible method is vacuum-drying, which allows effective removal of moisture under low pressure, temperature, and oxygen levels, and it is useful for thermolabile products [107].

Regarding the evaluation of the drying process of maqui and murta fruits, it was reported that the preservation techniques—freeze, convective, sun, infrared, and vacuum drying—result in a final maqui product with proper levels of phenolic compounds [38]. All these drying techniques showed a higher content of phenolic and antioxidant compounds, and freeze-dried samples retained over 60% of delphinidin and cyanidin derivatives of fresh fruits [38].

The convective hot air drying (40 to 80 °C) of maqui berries showed that a thermal load and not a high temperature are the main factors that affect the stability of bioactive compounds. At 40 °C, there was a long exposure of the berries to hot air compared to the drying process at 80 °C [39]. Above 60 °C, the bioactive components, such as β-carotene, tocopherols, anthocyanin, and vitamin B6, were not significantly affected, while gallic and ellagic acids increased, as a result of the conversion of hydrolyzable tannins. This phenomenon indicates that the loss of antioxidant activity is compensated for by a probable formation of bioactive components directly related to TPC [39]. Similar studies on murta berries (40 to 80 °C) showed that the β-carotene, total phenolic, and flavonoid contents show a significant decrease during the drying process compared to fresh fruit. However, the ORAC value showed similar antioxidant activity at higher drying temperatures (70–80 °C) compared to fresh fruit [108]. Otherwise, convective and combined convective-infrared conditions at 40, 50, and 60 °C and 400–800 W show that chromaticity coefficients a* and b*, the total surface color difference (ΔE), and TPC are dependent on the mode of heat supply. In addition, a constant temperature and high infrared power 40 °C/800 W reduced the drying time, resulting in dried samples with the highest TPC [42]. A comparative study conducted to evaluate the effect of convective hot-air drying at 65 and 80 °C and freeze drying of bioactive compounds of the Red Pearl-INIA variety of murta fruits showed that freeze-dried fruit retained higher values for TPC (21.924 mg g^{-1} DW) and TAC (0.134 mg g^{-1} DW) than the murta dried by convective hot-air at both temperatures, with a better retention of polyphenols and antioxidant activity during freeze drying [109].

3.2. Microencapsulation for Liquid Preparation

The anthocyanin content in maqui is significantly higher compared to other berries, which explains the great interest in its use for nutraceutical purposes. However, these bioactive compounds are highly labile, depending on the stabilization system used [103]. Microencapsulation technology can be used as a strategy to protect maqui anthocyanins in healthy food ingredients. Bastías-Montes et al., (2019) [110], showed that the microencapsulation of maqui can be one way to protect anthocyanins from degradation reactions and can be useful in liquid food preparation, such as for juice and yogurt, with a high content of bioactive compounds. The microencapsulation is a protective technological alternative through which certain bioactive substances in solid, liquid, or gas stage into microparticles with a diameter of 1–1000 µm, and has been widely used in the fields of medicine, cosmetics, food, textile, and advanced materials [111–113]. The unique advantage of microencapsulation lies in the fact that the core material is completely coated and isolated from the external environment. The aim is preserving them from various agents, as well as protecting them from oxidation reactions caused by light or oxygen.

Phenolic compounds are phytochemicals extensively metabolized after consumption; thus, the bioavailability should be considered when evaluating the potential health benefits of fruit ingestion. However, bioavailability is influenced by bioaccessibility, which is defined as the relative amount of nutrients or phytochemicals released from a complex food matrix in the lumen of the gastrointestinal tract, becoming available for absorption into the body [114,115]. The comparative analysis of microcapsules of maqui juice powdered by spray-drying or freeze-drying indicated that the morphology and particle size were the most relevant differences and affect the final solubility (70.4–59.5%) in water. However, no significant differences in the stability of anthocyanins in yogurt preparation and in the bioaccessibility after in vitro digestion were observed [104]. Other studies show that the encapsulation with inulin or sodium alginate allows maqui juice spray-drying until 133 days, and the highest encapsulation efficiency of anthocyanins was obtained with inulin. Both maqui juice microparticle methods improved the bioaccessibility (10%) of anthocyanins compared to maqui juice [116]. In murta, comparative studies showed that the highest bioactivity and storability of bioactive phenolics in juice extract were 28 ± 1 min for frozen-thawed fruits and 34 ± 1 min for fresh fruits [117]. In addition, the bioaccessibility index of polyphenols in fresh murta berries or their juice achieved a relatively high value (around 70%) at the end of the small intestine digestive step; however, the juice released the bioaccessible bioactive compounds in the earlier gastric stage, while the fresh fruit increased the release of bioactive compounds in the small intestine [117].

4. Healthy Potential of Patagonian berries

Phenolic compounds are effective antioxidants and can display various effects, including anti-microbial, anti-inflammatory, anti-mutagenic, anti-carcinogenic, anti-allergic, anti-platelet, vasodilator, and neuroprotective effects [65,67,118]. These properties have given rise to a new interest in finding plant species with a high phenolic content and relevant biological activity. The epidemiological evidence supporting the benefits of consuming a diet rich in foods containing polyphenols is strong [119–121]. In addition to the above, the richness of certain phenolic compounds present in different foods does not guarantee their absorption by the organism, which is how the bioavailability of each of them arises as one of the properties to study to correlate the intake and the effects thereof. The bioavailability appears to differ greatly among the various phenolic compounds, and the most abundant ones in our diet are not necessarily those that have the best bioavailability profile [121–124]. There has been a broad discussion about whether a high polyphenol content or high antioxidant activity can be associated with a real effect on human health. However, the results related to the preclinical evaluation of the antioxidant capacity and bioactivity of polyphenol extracts using cell cultures, isolated tissues, and animal models, before clinical trials, are still a good approach to understanding the healthy potential of several native fruits. In addition to the advances concerning characterization of the antioxidant capacity and the profile of bioactive molecules in fresh or processed Patagonian berries, advances have been made in the evaluation of the healthy potential of these berries (Figure 3). These sections summarize and discuss the literature regarding the progress in research on the effect of Patagonian fruit extracts in chronic diseases such as metabolic syndrome (MetS), diabetes, and cardiovascular diseases (CVD).

4.1. Polyphenols and Anti-Inflammatory Effects

Inflammation is a natural defense mechanism associated with many diseases, such as viral and microbial infections, allergies, obesity, and autoimmune and chronic diseases, and also includes reactions to an unhealthy diet or toxic compounds [120]. During the development of chronic diseases, and due to the higher production of reactive oxygen species (ROS), a series of oxidatives affect various proteins triggering the release of inflammatory signals that can lead to chronic inflammation [120,125]. Anti-inflammatory activity of polyphenols such as quercetin, rutin, morin, hesperetin, and hesperidin has been reported in both acute and chronic inflammation performed in animal models [120]. Polyphenols can exert anti-inflammatory effects by modulating enzymes involved in the metabolism of arginine and arachidonic acid, regulating cell activity, and influencing the production of proinflammatory molecules [120].

The high content of flavonoids, such as quercetin, present in arrayán and murta, suggests its participation as a protective agent in inflammatory diseases. Quercetin (also known as rutin), mainly present as quercetin 3-rutinoside in fruits and vegetables, is a flavonol described in the fruits of calafate, murta, and arrayán; a high concentration of quercetin in the methanolic extract obtained from the arrayán fruit has been observed [15,21]. Purified quercetin has a variety of biological effects, including antiallergic, anti-inflammatory, antioxidant, and platelet antiaggregant effects [126]. In addition, potential protective effects against acute lung injury (ALI) induced by endotoxin or lipopolysaccharide (LPS), a component of the cell wall of Gram-negative bacteria, have been described [127]. In mice, the previous administration of quercetin inhibits several mechanisms associated with the inflammatory process during pulmonary infection, such as the inhibition of arterial blood gas exchange induced by LPS and the infiltration of neutrophils in the lungs, suppression of LPS-induced expression of the macrophage inflammatory protein (MIP)-2, inactivation of matrix metalloproteinase (MMP)-9, and inhibition of Akt phosphorylation [127].

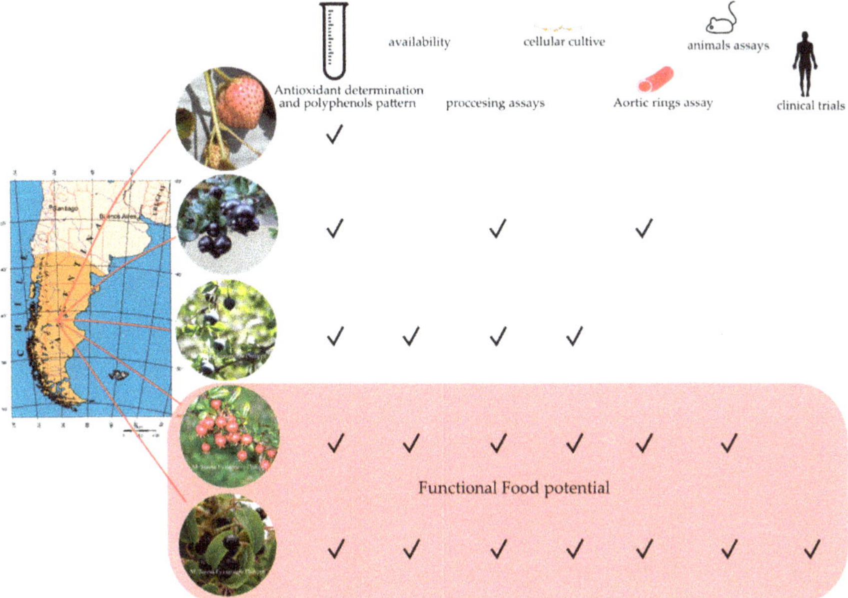

Figure 3. Summary of the Patagonian berries path to becoming functional foods. Maqui* is the native berry of Chile with major research progress concerning processing and the effect on chronic diseases. Murta* is the second most studied native berry, and two domesticated varieties are available in the market. Future studies are critical to strengthening the potential of arrayán**, calafate*, and Chilean strawberry** fruits. More details in the text. Photography credit to M. Teresa Eyzaguirre-Philippi (*) and Carlos R. Figueroa (**), map figure credit to commons.wikimedia.org/wiki/File:Pat_map.PNG, tube figure credit to https://thenounproject.com/term/test-tube/5544/, mouse figure credit to https://www.svgrepo.com/svg/53826/mouse, human figure credit to https://www.flaticon.com/free-icon/standing-human-body-silhouette_30473.

Studies conducted in animal models suggest that polyphenols in the diet have a positive effect on lung injury [128]. The inhalation of quercetin in radiation-induced pneumonitis in rats increases the number of leukocytes and erythrocytes in the blood, reduces the number of inflammatory cells in the bronchoalveolar lavage fluid, reduces hemorrhage and the infiltration of inflammatory cells, and suppresses the expression of proinflammatory cytokines that transform the growth factor β1 and interleukin-6 [128], suggesting that the inhalation of flavonols has the potential to become a new alternative in the treatment of lung diseases such as radiation pneumonitis.

As we previously stated, the maqui berry is the richest known natural source of delphinidins. An in vitro assay of this purified molecule showed an increase in the generation of nitrogen oxide (NO) in endothelial cells, decreased platelet adhesion, and anti-inflammatory effects. Additionally, it has been reported that delphinidins can counteract aging of the skin and inhibit osteoporosis [26]. Aqueous extracts of maqui berry prevent the oxidation of low-density lipoproteins (LDL) induced by copper, protect the cultures of human endothelial cells, and have anti-adipogenic and anti-inflammatory effects [75,78,129,130]. The extracts of maqui and calafate fruits have inhibitory properties of the inflammatory response generated by the interaction of adipocytes and macrophages [27]. These extracts showed a reduction of nitric oxide (NO) production, inhibition of the induction of nitric oxide synthase

(NOS) and TNF-alpha, and induction of the interleukin 10 (IL-10) gene expression; on this basis, it has been suggested that they could be potential therapeutic tools against the comorbidity associated with the development of obesity [27].

An in vitro assay performed in LPS-activated murine macrophage RAW-264 cells showed that extracts and subfractions of maqui berry, and also quercetin, gallic acid, luteolin, and myricetin, suppressed the LPS-induced production of NO, by downregulating iNOS and COX-2 expressions; according to the authors, the phenolic compounds anthocyanins, flavonoids, and organic acids, as the fractions, may provide a potential therapeutic tool for inflammation-associated disorders [131]. The antioxidant and anti-inflammatory effects of water extracts of maqui berry were tested in a mouse dermatitis model showing an increase of interferon-gamma (IFN-γ) levels and a decrease of interleukin-4 (IL-4), suggesting its potential use for atopic dermatitis treatment [132].

Studies in humans showed that anthocyanin maqui extract normalized H_2O_2 and IL-6 concentrations in exhaled breath condensates (EBC) by asymptomatic smokers [133], suggesting that the maqui could be considered an interesting alternative for dietary management in patients with respiratory disorders. Another study showed that the extracts of leaves and berries of murta have a strong anti-inflammatory activity when applied topically in mice, due to several pentacyclic triterpene acids, including the 2-a-hydroxy derivatives alphitolic, asiatic, and corosolic acids [129,134–136].

4.1.1. Polyphenols and Metabolic Syndrome

Metabolic syndrome (MetS) includes several metabolic abnormalities, such as abdominal obesity, hypertension, insulin resistance, and dyslipidemia. MetS has been associated with an increased risk of CVD and type 2 diabetes mellitus (T2DM) [67]. The onset and progression of MetS are mediated by body weight and blood pressure reduction, as well as improvement in insulin-sensitivity and lipid metabolism [119]. The beneficial effects of polyphenols, mainly flavonoids, are associated with their interaction with several molecular pathways involved in the metabolism of glucose and the regulation of insulin-signaling pathways [67].

A remarkable activity of polyphenols is their ability to retard carbohydrate digestion by the direct inhibition of enzymes, such as R-glucosidase and R-amylase [137]. As a result, the inhibition of these enzymes reduces the glucose absorption rate. It was reported that the crude extract of murta and maqui leaf rich in polyphenols—lavan-3-ol polymers, quercetin glucoside, and kaempferol glucoside—showed an effective inhibitory effect by a non-competitive mechanism on R-amylase and R-glucosidase f [137]. The above suggest a potential effect of these extracts in regulating postprandial hyperglycemia. In a murine model of type II diabetes, the oral administration of a standardized anthocyanin-rich formulation from maqui and pure delphinidin 3-sambubioside-5-glucoside (D3S5G) showed a dose-dependent decrease of fasting blood glucose levels and glucose production in rat liver cells [138].

A clinical trial conducted on individuals with a moderate glucose intolerance, daily supplemented with 180 mg Delphinol®, a standardized, water-soluble maqui berry extract, for three months, showed a progressive decrease of glycosylated hemoglobin, reduction of LDL and VLDL after one month, and increase of HDL from the baseline during the entire treatment period, without changes of total cholesterol and triglycerides, suggesting that longer treatment has a better effect on the glycemic and lipid profile [24,25]. A clinical pharmacokinetic study showed that after single-dose supplementation with Delphinol®, delphinidin-3-*O*-glucoside, and cyanidin-3-*O*-sambubioside, the selected anthocyanins in the assay reached the maximal concentration after approximately 1 and 2 h, respectively, confirming the bioavailability of these anthocyanins, and also their fast uptake and metabolism [139].

4.1.2. Cardiovascular Effects

CVD is the primary cause of mortality and morbidity worldwide. There is substantial evidence that early events of asymptomatic hyperglycemia increase the risk of CVD, even in the absence of diabetes [140]. Hyperglycemia is associated with endothelial dysfunction, characterized by reduced endothelium-dependent vasodilation (EDV), which is usually used as a measure to prove endothelial function in different pathological conditions [141].

Murta and arrayán berries might have beneficial effects on the management of cardiovascular diseases. The vasoprotective activity of the extract of these fruits could be associated with a cocktail of different molecules rather than a particular molecule. It was reported that a murta extract rich in gallic acid, catechin, quercetin-3-β-D-glucoside, myricetin, quercetin, and kaempferol did not generate toxic effects on human endothelial cells and had significant antioxidant activity against lipid peroxidation and superoxide anion and ROS production [142]. The same extract showed a dose-dependent vasodilator activity in aortic rings in the presence of endothelium, whose hypotensive mechanism is partially mediated by large conductance calcium-dependent potassium channels and nitric oxide synthase/guanylate cyclase [142]. Conversely, a methanolic extract of arrayán fruit harvested from a natural population located at Antuco (Biobio Region, Chile) containing quercetin-3-rutinoside, petunidin-3-arabinoside, peonidin-3-galactoside, malvidin-3-arabinoside, and peonidin-3-arabinoside arabinoside, showed vasoprotection properties [21]. Briefly, the methanolic extract of arrayán fruit showed dose-dependent (0.1, 1, and 10 mg/mL) protection of the acetylcholine-induced relaxation carried out in rat aortic rings (isolated from same litter animals) preincubated with a high level of glucose, a condition that drastically affects the endothelium-dependent relaxation induced by acetylcholine [21]. The above results suggest that the extract of Patagonian berries may act as a vasoprotector, which allows them to be projected as useful tools to prevent and treat diseases associated with vascular damage induced by high glucose levels (e.g., postprandial hyperglycemia) [143].

Patagonian fruits not only have a high content of polyphenolic compounds, but also have other vasoactive compounds. It was reported that the alkaloid 8-oxo-9-dihydromakomakine extracted from maqui leaves induced a dose-dependent relaxation of aortic rings precontracted with phenylephrine; the induced vasorelaxation was independent of endothelium and partially reduced plasma membrane depolarization-induced contraction, suggesting a protective effect of maqui alkaloids in the treatment of cardiovascular pathologies [144].

A clinical trial conducted in healthy, overweight, and smoker subjects showed that the daily consumption of anthocyanins was associated with reduced levels of oxidative damage markers in plasma (oxidized low-density lipoprotein; Ox-LDL) and urine (F2-isoprostanes). The values returned to the baseline value after 40 days of follow-up, and no significant differences were observed for anthropometric characteristics, ambulatory blood pressure, and the lipid profile [145].

5. Some Commercial Aspects

In Chile, maqui and murta are the primary Patagonian berries marketed, and most of them are exported for consumers worldwide (Figure 3). Concerning maqui, the principal harvest is from woodland shrubs. Although, according to the Center of Native plants of Chile (Universidad de Talca), this university published the applications of three domesticated varieties of maqui for their commercial use in the Official Gazette of Chile [146]. Romo and Bastías, (2016) [40] reported that, in 2016, there were 21 companies in Chile related to maqui commercialization since 2009. The Chilean market is focused on the preparation of beverages or juices based on maqui berry. Of these, 13 companies are located in the Metropolitan Region (62%), and the rest is distributed in the other regions, concentrated between the Maule Region and that of Araucanía [40]. In turn, maqui berry-based products can be found in the international market as frozen, juiced, dehydrated, canned, and other fruit preparations. During 2018, the maqui production in Chile included (i) 79,132 Kg of frozen fruit with a worth of US $ 598,207 and a mean value of 7.6 US/kg; and (ii) 3,870 Kg of drying fruit with a worth of US $ 105,269 and a mean value of 27.2 US/kg [147]. The main target markets were the USA, South Korea, Germany, and Japan. According to the Forest Institute (INFOR), 75% of maqui berries are exported freeze-dried [148].

In Chile, murta harvest is from woodland shrubs and domesticate varieties [41]. In 1996, the Agricultural Research Institute of Chile (INIA) developed a domestication program that began with the collection of wild germplasm [15,136]. This program included the development of protocols of plant multiplication [149,150], and the study of genetic diversity by molecular, phenotypic, and agronomic characterization of the wild germplasm [150]. According to a prospective study for new food introduction in the European Union requested by the Chilean Office of Agricultural Studies and Policies (OPEPA) during 2016, the exportation of principally fresh fruit was close to 3,000 Kg, with a worth of US $80,000. The major exportation markets were Italy, Korea, and France, among others [41]. No available information about the arrayán commercialization or breading program was found. However, some companies are interested in including some functional arrayán derivate products.

According to the novel food catalog of the European Union, maqui berry has an authorized use only as or in food supplements, and any other food uses have to be approved for the EU-Novel Food Regulation [151]. Regarding murta berry, the information currently available suggests that this fruit meets the requirements for the novel food solicitation [41].

Concerning Chilean native strawberry, no agro-industrial products have been generated, and this could be because production volumes are low enough to satisfy the demand for raw materials [43] (Figure 3). However, the "Slow Food" Foundation for Biodiversity, which promotes the protection of the biodiversity of food and its environmentally friendly production around the world, has incorporated the Chilean strawberry in the world project known as "The Ark of Taste" (Slow Food Foundation for Biodiversity, 2015) [152]. This critical tendency, associated with the rescue of gastronomic traditions and the growing market gourmet in Chile, can contribute to generating Chile's public policies regarding protection of the cultural and gastronomic heritage.

With regard to calafate, INIA coordinated the grant conducted for the generation of new varieties for a natural color generation. The project "Territorial Pole for the development of high value colorants and antioxidants for the food industry from highly dedicated raw materials produced in the south-central zone of Chile" includes the participation of INIA and agro-industrial companies and it is an initiative of "Territorial Poles of Strategic Development" created by the Foundation for Agrarian Innovation (FIA), with resources provided by the Strategic Investment Fund (FIE) [153].

6. Conclusions

This review provides relevant information about the native Patagonian berries—maqui, murta, calafate, arrayán, and Chilean strawberry—that could be used as a functional food due to its diverse and high flavonol and anthocyanin contents that can prevent inflammatory-, metabolic syndrome-, and cardiovascular-associated pathologies. Within the fruits discussed in this review, maqui is the native berry with major potential, followed by the murta fruit. Future functional studies and the production of cultivars are critical to strengthening the potential of these fruits in the food market.

Author Contributions: Writing—original draft preparation, L.F., C.R.F., M.V., and R.V.; writing—review and editing, L.F., C.R.F., M.V., and R.V.; funding acquisition, L.F., C.R.F., M.V., and R.V.

Funding: This research was funded by the National Commission for Scientific and Technological Research (CONICYT, Chile), grants number R17A10001 to L.F. and R.V.; FONDECYT/Regular 1181310 to C.R.F.

Acknowledgments: The authors thank M. Teresa Eyzaguirre-Philippi, president of "Fundación R.A. Philippi de Estudios Naturales" (http://fundacionphilippi.cl/), for her photographic material used in the figures of this review and highlight her contribution to the appreciation of the native flora of Chile.

Conflicts of Interest: The authors declare no conflict of interest. The founding sponsors had no role in the writing of the manuscript, and in the decision to publish this review.

References

1. Coronato, A.; Coronato, F.; Mazzoni, E.; Vázquez, M. *The physical geography of Patagonia and Tierra del Fuego. The Late Cenozoic of Patagonia and Tierra del Fuego*; Rabassa, J., Ed.; Elsevier: Amsterdam, The Netherlands, 2008; pp. 13–56.

2. Hoffmann, A.; Farga, C.; Lastra, J.; Veghazi, E. *Plantas Medicinales de Uso Comun en Chile*; Fundación Claudio Gay: Santiago de Chile, Chile, 1992; 257p.

3. Mösbach, E.W. *Botánica Indígena de Chile*; Museo Chileno de Arte Precolombino; Editorial Andrés Bello; Fundación Andes: Santiago de Chile, Chile, 1992; 140p.

4. Ladio, A.H.; Lozada, M. Patterns of use and knowledge of wild edible plants in distinct ecological environments: A case study of a Mapuche community from Nothwestern Patagonia. *Biodivers. Conserv.* **2004**, *13*, 1153–1173. [CrossRef]

5. Estomba, D.; Ladio, A.; Lozada, M. Medicinal wild plant knowledge and gathering patterns in a Mapuche community from North-western Patagonia. *J. Ethnopharmacol.* **2006**, *103*, 109–119. [CrossRef] [PubMed]

6. Díaz-Forestier, J.; León-Lobos, L.; Marticorena, A.; Celis-Diez, J.L.; Giovannini, P. Native Useful Plants of Chile: A Review and Use Patterns. *Econ. Bot.* **2019**, *73*, 112–126. [CrossRef]

7. Barreau, A.; Ibarra, J.T.; Wyndham, F.S.; Rojas, A.; Kozak, R.A. How can we teach our children if we cannot access the forest? Generational change in mapuche knowledge of wild edible plants in Andean temperature ecosystems of Chile. *J. Ethnobiol.* **2016**, *36*, 412–432. [CrossRef]

8. Rivera, D.; Verde, A.; Fajardo, J.; Inocencio, C.; Obon, C.; Heinrich, M. *Guia Etnobotanica de los Alimentos Locales Recolectados en la Provincia de Albacete*; Instituto de Estudios Albacetenses; Diputación de Albacete: Albacete, Spain, 2006.

9. Egea, I.; Sánchez-Bel, P.; Romojaro, F.; Pretel, M.T. Six edible wild fruits as potential antioxidant additives or nutritional supplements. *Plant Foods Hum. Nutr.* **2010**, *65*, 121–129. [CrossRef] [PubMed]

10. Pereira, M.C.; Steffens, R.S.; Jablonski, A.; Hertz, P.F.; Rios Ade, O.; Vizzotto, M.; Flôres, S.H. Characterization and antioxidant potential of Brazilian fruits from the Myrtaceae family. *J. Agric. Food Chem.* **2012**, *60*, 3061–3067. [CrossRef]

11. Speisky, H.; López Alarcón, C.; Gómez, M.; Fuentes, J.; Sandoval Acuña, C. First web-based database on total phenolics and oxygen radical absorbance capacity (ORAC) of fruits produced and consumed within the South Andes Region of South America. *J. Agric. Food Chem.* **2012**, *60*, 8851–8859. [CrossRef]

12. Ramos, A.S.; Souza, R.O.S.; Boleti, A.P.A.; Bruginski, E.R.D.; Lima, E.S.; Campos, F.R.; Machado, M.B. Chemical characterization and antioxidant capacity of the araçá-pera (Psidium acutangulum): An exotic Amazon fruit. *Food Res. Int.* **2015**, *75*, 315–327. [CrossRef]

13. Kongkachuichai, R.; Charoensiri, R.; Yakoh, K.; Kringkasemsee, A.; Insung, P. Nutrients value and antioxidant content of indigenous vegetables from Southern Thailand. *Food Chem.* **2015**, *173*, 838–846. [CrossRef]

14. Barros, R.G.C.; Andrade, J.K.S.; Denadai, M.; Nunes, M.L.; Narain, N. Evaluation of bioactive compounds potential and antioxidant activity in some Brazilian exotic fruit residues. *Food Res. Int.* **2017**, *102*, 84–92. [CrossRef]

15. Ulloa-Inostroza, E.M.; Ulloa-Inostroza, E.G.; Alberdi, M.; Peña-Sanhueza, D.; González-Villagra, J.; Jaakola, L.; Reyes-Díaz, M. Native Chilean Fruits and the Effects of their Functional Compounds on Human Health, Superfood and Functional Food Viduranga Waisundara. Available online: https://www.intechopen.com/books/superfood-and-functional-food-an-overview-of-their-processing-and-utilization/native-chilean-fruits-and-the-effects-of-their-functional-compounds-on-human-health (accessed on 29 June 2019).

16. Schmeda-Hirschmann, G.; Jiménez-Aspee, F.; Theoduloz, C.; Ladio, A. Patagonian berries as native food and medicine. *J. Ethnopharmacol.* **2019**, *241*, 111979. [CrossRef] [PubMed]

17. Simirgiotis, M.J.; Theoduloz, C.; Caligari, P.D.S.; Schmeda-Hirschmann, G. Comparison of phenolic composition and antioxidant properties of two native Chilean and one domestic strawberry genotypes. *Food Chem.* **2009**, *113*, 377–385. [CrossRef]

18. Simirgiotis, M.J.; Bórquez, J.; Schmeda-Hirschmann, G. Antioxidant capacity, polyphenol content and tandem HPLCDAD- ESI/MS profiling of phenolic compounds from the South American berries *Luma apiculata* and *L. chequen. Food Chem.* **2013**, *139*, 289–299. [CrossRef] [PubMed]

19. Brito, A.; Areche, C.; Sepúlveda, B.; Kennelly, E.; Simirgiotis, M. Anthocyanin characterization, total phenolic quantification and antioxidant features of some Chilean edible berry extracts. *Molecules* **2014**, *19*, 10936–10955. [CrossRef] [PubMed]

20. Ruiz, A.; Hermosín, I.; Mardones, C.; Vergara, C.; Herlitz, C.; Vega, M.; Dorau, C.; Winterhalter, P.; Von Baer, D. Polyphenols and antioxidant activity of calafate (Berberis microphylla) fruits and other native berries from southern Chile. *J. Agric. Food Chem.* **2010**, *58*, 6081–6089. [CrossRef] [PubMed]

21. Fuentes, L.; Valdenegro, M.; Gómez, M.G.; Ayala-Raso, A.; Quiroga, E.; Martínez, J.P.; Vinet, R.; Caballero, E.; Figueroa, C.R. Characterization of fruit development and potential health benefits of arrayan (*Luma apiculata*), a native berry of South America. *Food Chem.* **2016**, *196*, 1239–1247. [CrossRef] [PubMed]

22. Rozzi, R.; Massardo, F. Las Plantas Medicinales Chilenas II y III. *Informe Nueva Medicina* **1994**, *1*, 16–17.

23. Massardo, F.; Rozzi, R. Usos medicinales de la flora nativa chilena. *Ambiente y Desarrollo* **1996**, *12*, 76–81.

24. Alvarado, J.; Schoenlau, F.; Leschot, A.; Salgad, A.M.; Vigil Portales, P. Delphinol®standardized maqui berry extract significantly lowers blood glucose and improves blood lipid profile in prediabetic individuals in three-month clinical trial. *Panminerva Med.* **2016**, *58* (Suppl. 1), 1–6.

25. Alvarado, J.; Leschot, A.; Olivera Nappa, Á.; Salgado, A.; Rioseco, H.; Lyon, C.; Vigil, P. Delphinidin-Rich Maqui Berry Extract (Delphinol (R)) Lowers Fasting and Postprandial Glycemia and Insulinemia in Prediabetic Individuals during Oral Glucose Tolerance Tests. *BioMed Res. Int.* **2016**, *2016b*, 9070537. [CrossRef]

26. Watson, R.R.; Schönlau, F. Nutraceutical and antioxidant effects of a delphinidin-rich maqui berry extract Delphinol®: A review. *Minerva Cardioangiol.* **2015**, *63* (Suppl. 1), 1–12. [PubMed]

27. Reyes-Farias, M.; Vasquez, K.; Ovalle-Marin, A.; Fuentes, F.; Parra, C.; Quitral, V.; Jimenez, P.; Garcia-Diaz, D.F. Chilean native fruit extracts inhibit inflammation linked to the pathogenic interaction between adipocytes and macrophages. *J. Med. Food* **2015**, *18*, 601–608. [CrossRef] [PubMed]

28. Barrett, D.M.; Beaulieu, J.C.; Shewfelt, R. Color, flavor, texture, and nutritional quality of fresh-cut fruits and vegetables: Desirable levels, instrumental and sensory measurement, and the effects of processing. *Crit. Rev. Food Sci. Nutr.* **2010**, *50*, 369–389. [CrossRef] [PubMed]

29. Kramer, A. Evaluation of quality of fruits and vegetables. In *Food Quality*; Irving, G.W., Jr., Hoover, S.R., Eds.; American Association for the Advancement of Science: Washington, DC, USA, 1965; pp. 9–18.

30. Fuentes, L.; Figueroa, C.R.; Valdenegro, M. Recent Advances in Hormonal Regulation and Cross-Talk during Non-Climacteric Fruit Development and Ripening. *Horticulturae* **2019**, *5*, 45. [CrossRef]

31. Figueroa, C.R.; Pimentel, P.; Gaete-Eastman, C.; Moya, M.; Herrera, R.; Caligari, P.D.; Moya-León, M.A. Softening rate of the Chilean strawberry (Fragaria chiloensis) fruit reflects the expression of polygalacturonase and pectate lyase genes. *Postharv. Biol. Technol.* **2008**, *49*, 210–220. [CrossRef]

32. Concha, C.M.; Figueroa, N.E.; Poblete, L.A.; Oñate, F.A.; Schwab, W.; Figueroa, C.R. Methyl jasmonate treatment induces changes in fruit ripening by modifying the expression of several ripening genes in Fragaria chiloensis fruit. *Plant Physiol. Biochem.* **2013**, *70*, 433–444. [CrossRef] [PubMed]

33. Garrido-Bigotes, A.; Figueroa, P.M.; Figueroa, C.R. Jasmonate metabolism and its relationship with abscisic acid during strawberry fruit development and ripening. *J. Plant Growth Regul.* **2018**, *37*, 101–113. [CrossRef]

34. Fredes, C.; Montenegro, G.; Zoffoli, J.P.; Gómez, M.; Robert, P. Polyphenol content and antioxidant activity of maqui (*Aristotelia chilensis* Molina Stuntz) during fruit development and maturation in Central Chile. *Chil. J. Agric. Res.* **2012**, *72*, 582–589. [CrossRef]

35. Rodriguez, R.; Marticorena, C.; Alarcón, D.; Baeza, C.; Cavieres, L.; Finot, V.L.; Fuentes, N.; Kiessling, A.; Mihoc, M.; Pauchard, A.; et al. Catálogo de las plantas vasculares de Chile. *Gayana Botánica* **2018**, *75*, 1–430. [CrossRef]

36. Gomes, F.C.; Lacerda, I.C.; Libkind, D.; Lopes, C.; Carvajal, E.J.; Rosa, C.A. Traditional foods and beverages from South America: Microbial communities and production strategies. *Ind. Ferment. Food Process. Nutr. Sour. Prod. Strateg.* **2009**, *3*, 79–114.

37. Hoffmann, A.E. *Flora Silvestre de Chile*, 5th ed.; Zona Araucana Edicione; Fundación Claudio Gay: Santiago de Chile, Chile, 2005; 257p.

38. Quispe-Fuentes, I.; Vega-Gálvez, A.; Aranda, M. Evaluation of phenolic profiles and antioxidant capacity of maqui (*Aristotelia chilensis*) berries and their relationships to drying methods. *J. Sci. Food Agric.* **2018**, *98*, 4168–4176. [CrossRef] [PubMed]

39. Rodríguez, K.; Ah-Hen, K.S.; Vega-Gálvez, A.; Vásquez, V.; Quispe-Fuentes, I.; Rojas, P.; Lemus-Mondaca, R. Changes in bioactive components and antioxidant capacity of maqui, *Aristotelia chilensis* [Mol] Stuntz, berries during drying. *LWT Food Sci. Technol.* **2016**, *65*, 537–542. [CrossRef]

40. Romo, R.; Bastías, J.M. Estudio de Mercado del Maqui. PYT-0215-0219. Perspectiva del Mercado Internacional para el Desarrollo de la Industria del Maqui: Un Análisis de las Empresas en Chile. Available online: http://bibliotecadigital.fia.cl/bitstream/handle/20.500.11944/146064/Informe%20Final%20Estudio%20de%20Mercado%20MAQUI.pdf?sequence=1&isAllowed=y (accessed on 28 May 2019).

41. Estudio Preparación de Expedientes Técnicos para la Presentación y Solicitud de Autorización de Alimentos Nuevos o Tradicionales de Terceros Países para Exportar a la Unión Europea. Available online: https://www.odepa.gob.cl/wp-content/uploads/2017/12/Informe-Estudio-Novel-Foods-y-anexos-1.pdf (accessed on 28 May 2019).

42. Puente-Díaz, L.K.; Ah-Hen, A.; Vega-Gálvez, R.; Lemus-Mondaca; Di Scala, K. Combined infrared-convective drying of murta (Ugni molinae Turcz.) berries: Kinetic modeling and quality assessment. *Dry Technol.* **2013**, *31*, 329–338.

43. Figueroa, C.R.; Concha, C.M.; Figueroa, N.E.; Tapia, G. Frutilla Chilena Nativa Fragaria chiloensis. Available online: http://www.procisur.org.uy/adjuntos/13d4953eae2f_Chilena_PROCISUR.pdf (accessed on 28 May 2019).

44. Lavin, A.; Vega, A. Caracterizacion de frutos de murtilla (*Ugni molinae* Turcz.) en el area de Cauquenes. *Agricultura Tecnica* **1996**, *56*, 64–67.

45. Arena, M.E.; Curvetto, N. *Berberis buxifolia* fruiting: Kinetic growth behavior and evolution of chemical properties during the fruiting period and different growing seasons. *Scientia Horticulturae* **2008**, *118*, 120–127. [CrossRef]

46. Stevens, M.A.; Kader, A.A.; Albright, M. Potential for increasing tomato flavor via increased sugar and acid content. *J. Am. Soc. Horticult. Sci.* **1979**, *104*, 40–42.

47. Batista-Silva, W.; Nascimento, V.L.; Medeiros, D.B.; Nunes-Nesi, A.; Ribeiro, D.M.; Zsögön, A.; Araújo, W.L. Modifications in Organic Acid Profiles During Fruit Development and Ripening: Correlation or Causation? *Front Plant. Sci.* **2018**, *9*, 1689. [CrossRef] [PubMed]

48. Cherian, S.; Figueroa, C.R.; Nair, H. 'Movers and shakers' in the regulation of fruit ripening: A cross-dissection of climacteric versus non-climacteric fruit. *J. Exp. Bot.* **2014**, *65*, 4705–4722. [CrossRef] [PubMed]

49. Arena, M.E.; Postemsky, P.; Curvetto, N.R. Accumulation patterns of phenolic compounds during fruit growth and ripening of *Berberis buxifolia*, a native Patagonian species. *N. Z. J. Bot.* **2012**, *50*, 1528. [CrossRef]

50. Arena, M.E.; Zuleta ADyner, L.; Constenla DCeci, L.; Curvetto, N.R. *Berberis buxifolia* fruit growth and ripening: Evolution in carbohydrate and organic acid contents. *Scientia Horticulturae* **2013**, *158*, 52–58. [CrossRef]

51. Gonzalez, M.; Gaete-Eastman, C.; Valdenegro, M.; Figueroa, C.R.; Fuentes, L.; Herrera, R.; Moya-León, M.A. Aroma development during ripening of Fragaria chiloensis fruit and participation of an alcohol acyltransferase (FcAAT1) gene. *J. Agric. Food Chem.* **2009**, *57*, 9123–9132. [CrossRef] [PubMed]

52. Scheuermann, E.; Seguel, I.; Montenegro, A.; Bustos ROHormazabal, E.; Quiroz, A. Evolution of aroma compounds of murtilla fruits (*Ugni molinae* Turcz) during storage. *J. Sci. Food Agric.* **2008**, *88*, 485–492. [CrossRef]

53. Hummer, K.E.; Hancock, J. Strawberry Genomics: Botanical History, Cultivation, Traditional Breeding, and New Technologies. In *Genetics and Genomics of Rosaceae. Plant Genetics and Genomics: Crops and Models*; Folta, K.M., Gardiner, S.E., Eds.; Springer: New York, NY, USA, 2009; Volume 6.

54. Chen, L.; Opara, U.L. Texture measurement approaches in fresh and processed foods—A review. *Food Res. Int.* **2013**, *51*, 823–835. [CrossRef]

55. Figueroa, C.R.; Rosli, H.G.; Civello, P.M.; Martínez, G.A.; Herrera, R.; Moya-León, M.A. Changes in cell wall polysaccharides and cell wall degrading enzymes during ripening of Fragaria chiloensis and Fragaria× ananassa fruits. *Scientia Horticulturae* **2010**, *124*, 454–462. [CrossRef]

56. Valdenegro, M.; Fuentes, L.; Urtubia, A.; Herrera, R.; Moya-Leon, M.A. Modified atmosphere packaging preserves quality and antioxidant properties in Fragaria chiloensis. *J. Biotechnol.* **2010**, *150* (Suppl. 1), S335.

57. Vicente, A.R.; Ortugno, C.; Rosli, H.; Powell, A.L.; Greve, L.C.; Labavitch, J.M. Temporal sequence of cell wall disassembly events in developing fruits. 2. Analysis of blueberry (*Vaccinium* species). *J. Agric. Food Chem.* **2007**, *55*, 4125–4130. [CrossRef]

58. Defilippi, B.; Robledo, P.; Becerra, C.; Seguel, I. Comportamiento en postcosecha de dos variedades de murtilla: South Pearl INIA y Red Pearl INIA. Available online: http://www.inia.cl/postcosecha/files/2015/02/Defilippi_murtilla_Iquique.pdf (accessed on 26 June 2019).

59. Robledo, P.; Soto-Alvear, S.; Seguel, I.; Defilippi, B. Evaluación de Tecnologías de Postcosecha en Murtilla. Available online: http://www.inia.cl/postcosecha/files/2015/02/Poster_murtilla2012.pdf (accessed on 25 May 2019).

60. Dixon, R.A.; Paiva, N.L. Stress-induced phenylpropanoid metabolism. *Plant Cell* **1995**, *7*, 1085. [CrossRef] [PubMed]

61. Landrum, L.; Donoso, C. *Ugni molinae* (Myrtaceae), a potential fruit crop for regions of Mediterranean, maritime and subtropical climates. *Econ. Bot.* **1990**, *44*, 536–539.

62. Rice-Evans, C.; Miller, N.J.; Paganga, G. Structure antioxidant activity relationships of flavonoids and phenolic acids. *Free Radic. Biol. Med.* **1996**, *20*, 933–956. [CrossRef]

63. Rice-Evans, C.; Miller, N.; Paganga, G. Antioxidant properties of phenolic compounds. *Trends Plant Sci.* **1997**, *2*, 152–159. [CrossRef]

64. Soto-Vaca, A.; Gutierrez, A.; Losso, J.N.; Xu, Z.; Finley, J.W. Evolution of phenolic compounds from color and flavor problems to health benefits. *J. Agric. Food Chem.* **2012**, *60*, 6658–6677. [CrossRef] [PubMed]

65. Scalbert, A.; Manach, C.; Morand, C.; Remesy, C. Dietary polyphenols and the prevention of diseases. *Crit. Rev. Food Sci. Nutr.* **2005**, *45*, 287–306. [CrossRef] [PubMed]

66. Dai, J.; Mumper, R.J. Plant phenolics: Extraction, analysis and their antioxidant and anticancer properties. *Molecules* **2010**, *15*, 7313–7352. [CrossRef] [PubMed]

67. Russo, B.; Picconi, F.; Malandrucco, I.; Frontoni, S. Flavonoids and Insulin-Resistance: From Molecular Evidences to Clinical Trials. *Int. J. Mol. Sci.* **2019**, *20*, 2061. [CrossRef] [PubMed]

68. Wikimedia Commons. Available online: https://commons.wikimedia.org/wiki/Main_Page (accessed on 26 May 2019).

69. Portal Antioxidantes. Available online: http://www.portalantioxidantes.com/orac-base-de-datos-actividad-antioxidante-y-contenido-de-polifenoles-totales-en-frutas/ (accessed on 28 May 2019).

70. Genskowsky, E.; Puente, L.A.; Pérez-Álvarez, J.A.; Fernández-López, J.; Muñoz, L.A.; Viuda-Martos, M. Determination of polyphenolic profile, antioxidant activity and antibacterial properties of maqui [Aristotelia chilensis (Molina) Stuntz] a Chilean blackberry. *J. Sci. Food Agric.* **2016**, *96*, 4235–4242. [CrossRef] [PubMed]

71. Escribano-Bailón, M.; Alcalde-Eon, C.; Muñoz, O.; Rivas-Gonzalo, J.; Santos-Buelga, C. Anthocyanins in berries of maqui (Aristotelia chilensis (Mol.) Stuntz). *Phytochem. Anal.* **2006**, *17*, 8–14. [CrossRef]

72. Junqueira-Gonçalves, M.P.; Yáñez, L.; Morales, C.; Navarro, M.; AContreras, R.; Zúñiga, G.E. Isolation and characterization of phenolic compounds and anthocyanins from Murta (Ugni molinae Turcz.) fruits. Assessment of antioxidant and antibacterial activity. *Molecules* **2015**, *20*, 5698–5713. [CrossRef]

73. Dávalos, A.; Gómez-Cordovés, C.; Bartolomé, B. Extending applicability of the oxygen radical absorbance capacity (ORAC—fluorescein) assay. *J. Agric. Food Chem.* **2004**, *52*, 48–54. [CrossRef]

74. Fredes, C.; Montenegro, G.; Zoffoli, J.P.; Santander, F.; Robert, P. Comparison of the total phenolic content, total anthocyanin content and antioxidant activity of polyphenol-rich fruits grown in Chile. *Cienc. Investig. Agrar.* **2014**, *41*, 49–60. [CrossRef]

75. Céspedes, C.; El-Hafidi, M.; Pavon, N.; Alarcon, J. Antioxidant and cardioprotective activities of phenolic extracts from fruits of Chilean blackberry Aristotelia chilenesis (Elaeocarpaceae), Maqui. *Food Chem.* **2008**, *108*, 820–829. [CrossRef]

76. Mariangel, E.; Reyes-Diaz, M.; Lobos, W.; Bensch, E.; Schalchli, H.; Ibarra, P. The antioxidant properties of calafate (Berberis microphylla) fruits from four different locations in southern Chile. *Ciencia e Investigación Agraria* **2013**, *40*, 161–170. [CrossRef]

77. González, B.; Vogel, H.; Razmilic, I.; Wolfram, E. Polyphenol, anthocyanin and antioxidant content in different parts of maqui fruits (*Aristotelia chilensis*) during ripening and conservation treatments after harvest. *Ind. Crops Prod.* **2015**, *76*, 158–165. [CrossRef]

78. Miranda-Rottmann, S.; Aspillaga, A.A.; Pérez, D.D.; Vasquez, L.; Martinez, A.L.F.; Leighton, F. Juice and phenolic fractions of the berry Aristotelia chilensis inhibit LDL oxidation in vitro and protect human endothelial cells against oxidative stress. *J. Agric. Food Chem.* **2002**, *50*, 7542–7547. [CrossRef] [PubMed]

79. He, K.; Valcic, S.; Timmermann, B.N.; Montenegro, G. Indole alkaloids from *Aristotelia chilensis* (Mol.) Stuntz. *Int. J. Pharm.* **1997**, *35*, 215–217. [CrossRef]

80. Ruiz, A.; Hermosín-Gutiérrez, I.; Vergara, C.; Von Baer, D.; Zapata, M.; Hitschfeld, A.; Obando, L.; Mardones, C. Anthocyanin profiles in south Patagonian wild berries by HPLC-DADESI- MS/MS. *Food Res. Int.* **2013**, *51*, 706–713. [CrossRef]

81. Fredes, C.; Yousef, G.G.; Robert, P.; Grace, M.H.; Lila, M.A.; Gómez, M.; Gebauer, M.; Montenegro, G. Anthocyanin profiling of wild maqui berries (*Aristotelia chilensis* [Mol.] Stuntz) from different geographical regions in Chile. *J. Sci. Food Agric.* **2014**, *94*, 2639–2648. [CrossRef] [PubMed]

82. Ramirez, J.E.; Zambrano, R.; Sepúlveda, B.; Kennelly, E.J.; Simirgiotis, M.J. Anthocyanins and antioxidant capacities of six Chilean berries by HPLC-HR-ESI-ToF-MS. *Food Chem.* **2015**, *176*, 106–114. [CrossRef]

83. Mahdavi, S.A.; Jafari, S.M.; Ghorbani, M.; Assadpoor, E. Spray-drying microencapsulation of anthocyanins by natural biopolymers: A review. *Drying Technol.* **2014**, *32*, 509–518. [CrossRef]

84. Gómez-Guillén, M.C.; Ihl, M.; Bifani, V.; Silva, A.; Montero, P. Edible films made from tuna-fish gelatin with antioxidant extracts of two different murta ecotypes leaves (Ugni molinae Turcz.). *Food Hydrocoll.* **2007**, *21*, 1133–1143. [CrossRef]

85. Barbosa-Cánovas, G.V.; Vega-Mercado, H. *Dehydration of Foods*; Chapman & Hall: New York, NY, USA, 1996.

86. Doymaz, I. Effect of citric acid and blanching pre-treatments on drying and rehydration of Amasya red apples. *Food Bioprod. Process.* **2010**, *88*, 124–132. [CrossRef]

87. Galaz, P.; Valdenegro, M.; Ramírez, C.; Nuñez, H.; Almonacid, S.; Simpson, R. Effect of drum drying temperature on drying kinetic and polyphenol contents in pomegranate peel. *J. Food Eng.* **2017**, *208*, 19–27. [CrossRef]

88. Valdenegro, M.; Almonacid, S.; Henríquez, C.; Lutz, M.; Fuentes, L.; Simpson, R. The Effects of Drying Processes on Organoleptic Characteristics and the Health Quality of Food Ingredients Obtained from Goldenberry Fruits (Physalis peruviana). Available online: https://www.omicsonline.org/scientific-reports/srep642.php (accessed on 28 May 2019).

89. Dev, S.R.S.; Padmini, T.; Adedeji, A.; Garie'py, Y.; Raghavan, G.S.V. A comparative study on the effect of chemical, microwave, and pulsed electric pretreatments on convective drying and quality of raisins. *Drying Technol.* **2008**, *26*, 1238–1243. [CrossRef]

90. Álvarez, C.A.; Aguerre, R.; Gomez, S.; Vidales, S.; Alzamora, S.M.; Gerschenson, L.N. Effect of blanching and glucose dipping pre-treatment on air-drying behavior of strawberries. In *Symposium on the Properties of Water. Proceedings of the Poster Session, Universidad de las Américas, México*; Argaiz, A., Lopez-Malo, A., Palou, E., Corte, P., Eds.; Universidad de las Americas: Puebla, Mexico, 1994; pp. 11–14.

91. Alzamora, S.M.; Greschenson, L.M.; Vidales, S.L.; Nieto, A. Structural changes in the minimal processing of fruits; some effects of blanching and sugar impregnation. In *Food Engineering 2000*; Fito, P., Ortega-Rodriguez, E., Barbosa-Cánovas, G.V., Eds.; Chapman & Hall: New York, NY, USA, 1996; pp. 117–139.

92. Carranza-Concha, J.; Benlloch, M.; Camacho, M.M.; Martínez-Navarrete, N. Effects of drying and pretreatment on the nutritional and functional quality of raisins. *Food Bioprod. Process.* **2012**, *90*, 243–248. [CrossRef]

93. Cinquanta, L.; Di Matteo, M.; Esti, M. Physical pretreatment of plums (Prunus domestica) Part 2. Effect on the quality characteristics of different prune cultivars. *Food Chem.* **2002**, *79*, 233–238. [CrossRef]

94. Di Matteo, M.; Cinquanta, L.; Galiero, G.; Crescitelli, S. Effect of a novel physical pretreatment process on the drying kinetics of seedless grapes. *J. Food Eng.* **2000**, *46*, 83–89. [CrossRef]

95. Doymaz, I. Influence of pretreatment solution on the drying of sour cherry. *J. Food Eng.* **2007**, *78*, 591–596. [CrossRef]

96. Azoubel, P.M.; Murr, F.E.X. Effect of pretreatment on the drying kinetics of cherry tomato (Lycopersicon esculentum var. cerasiforme). In *Transport Phenomena in Food Processing*; Welti-Chanes, J., Velez-Ruiz, F., Barbosa-Cánovas, G.V., Eds.; CRC Press: New York, NY, USA, 2003; pp. 137–151.

97. Feng, H.; Tang, J.; Mattinson, D.S.; Fellman, J.K. Microwave and spouted bed drying of frozen blueberries: The effect of drying and pre-treatment methods on physical properties and retention of flavour volatiles. *J. Food Process. Preserv.* **1999**, *23*, 463–479. [CrossRef]

98. Grabowski, S.; Marcotte, M. Pretreatment efficiency in osmotic dehydration of cranberries. In *Transport Phenomena in Food Processing*; Welti-Chanes, J., Velez-Ruiz, F., Barbosa-Cánovas, G.V., Eds.; CRC Press: New York, NY, USA, 2003; pp. 83–94.

99. Grabowski, S.; Marcotte, M.; Poirier, M.; Kudra, T. Drying Characteristics of Osmotically Pretreated Cranberries: Energy and Quality Aspects. 2002. Available online: http://cetcvarennes.nrcan.gc.ca/eng/publication/r2002083e.html (accessed on 12 June 2003).

100. Yang, P.A.P.; Wills, C.; Yang, C.S.T. Use of a combination process of osmotic dehydration and freeze drying to produce a raisin-type lowbush blueberry product. *J. Food Sci.* **1987**, *52*, 1651–1653. [CrossRef]

101. George, S.T.; Cenkowski, S.; Muir, W.E. A review of drying technologies for the preservation of nutritional compounds in waxy skinned fruit. In Proceedings of the 2004 North Central ASAE/CSAE Conference, Winnipeg, MB, Canada, 24–25 September 2004.

102. Desai, K.; Park, H. Recent developments in microencapsulation of food ingredients. *Drying Technol.* **2005**, *23*, 1361–1394. [CrossRef]

103. Robert, P.; García, P.; Fredes, C. Drying and preservation of polyphenols. In *Advances in Technologies for Producing Food-Relevant Polyphenols*; Cuevas-Valenzuela, J., Vergara Salinas, J.R., Pérez-Correa, J.R., Eds.; CRC Press, Taylor and Francis Group: Boca Raton, FL, USA, 2017; pp. 281–302.

104. Fredes, C.; Becerra, C.; Parada, J.; Robert, P. The Microencapsulation of Maqui (Aristotelia chilensis (Mol.) Stuntz) Juice by Spray-Drying and Freeze-Drying Produces Powders with Similar Anthocyanin Stability and Bioaccessibility. *Molecules* **2018**, *23*, 1227. [CrossRef]

105. Parra, H. Revision: Food microencapsulation. *Mag. Natl. Fac. Agron.* **2010**, *63*, 5669–5684.

106. Laokuldilok, T.; Kanha, N. Effects of processing conditions on powder properties of black glutinous rice (Oryza sativa L.) bran anthocyanins produced by spray drying and freeze drying. *LWT Food Sci. Technol.* **2015**, *64*, 405–411. [CrossRef]

107. Ah-Hen, K.S.; Zambra, C.E.; Agüero, J.E.; Vega-Gálvez, A.; Lemus-Mondaca, R. Moisture diffusivity coefficient and convective drying modelling of murta (*Ugni molinae* Turcz.): Influence of temperature and vacuum on drying kinetics. *Food Bioprocess. Technol.* **2013**, *6*, 919–930. [CrossRef]

108. Rodríguez, K.; Ah-Hen, K.; Vega-Gálvez, A.; López, J.; Quispe-Fuentes, I.; Lemus-Mondaca, R.; Galvez-Ranilla, L. Changes in bioactive compounds and antioxidant activity during convective drying of murta (*Ugni molinae* T.) berries. *Int. J. Food Sci. Technol.* **2014**, *49*, 990–1000. [CrossRef]

109. Alfaro, S.; Mutis, A.; Quiroz, A.; Seguel, I.; Scheuermann, E. Effects of drying techniques on murtilla fruit polyphenols and antioxidant activity. *J. Food Res.* **2014**, *3*, 73–82. [CrossRef]

110. Bastias-Montes, J.M.; Alarcón-Enos, J.; Quevedo-León, R.; Muñoz-Fariña, O.; Vidal-San-Martín, C. Effect of spray drying at 150, 160, and 170 °C on the physical and chemical properties of maqui extract (*Aristotelia chilensis* (Molina) Stuntz). *Chil. J. Agric. Res.* **2019**, *79*, 144–152. [CrossRef]

111. Pasin, B.; González, C.; Maestro, A. Microencapsulación con alginato en alimentos. Técnicas y aplicaciones. *Cienc Tecnol Alimentos* **2012**, *3*, 130–151.

112. Dubey, R.; Shami, T.C.; Bhasker Rao, K.U. Microencapsulation Technology and Applications. *Def. Sci. J.* **2009**, *59*, 82–95.

113. Saénz, C.; Tapia, S.; Chávez, J.; Robert, P. Microencapsulation by spray drying of bioactive compounds from cactus pear (*Opuntia ficus-indica*). *Food Chem.* **2009**, *114*, 616–622. [CrossRef]

114. Kamiloglu, S.; Pasli, A.A.; Ozcelik, B.; Capanoglu, E. Evaluating the in vitro bioaccessibility of phenolics and antioxidant activity during consumption of dried fruits with nuts. *LWT Food Sci. Technol.* **2014**, *56*, 284–289. [CrossRef]

115. Mandalari, G.; Bisignano, C.; Filocamo, A.; Chessa, S.; Sarò, M.; Torre, G.; Dugo, P. Bioaccessibility of pistachio (*Pistacia vera* L.) polyphenols, xanthophylls and tocopherols during simulated human digestion. *Nutrition* **2013**, *29*, 338–344. [CrossRef] [PubMed]

116. Fredes, C.; Osorio, M.J.; Parada, J.; Robert, P. Stability and bioaccessibility of anthocyanins from maqui (*Aristotelia chilensis* [Mol.] Stuntz) juice microparticles. *LWT Food Sci. Technol.* **2018**, *91*, 549–556. [CrossRef]

117. Ah-Hen, K.S.; Mathias-Rettig, K.; Gómez-Pérez, L.S.; Riquelme-Asenjo, G.; Lemus-Mondaca, R.; Muñoz-Fariña, O. Bioaccessibility of bioactive compounds and antioxidant activity in murta (*Ugni molinae* T.) berries juices. *Food Meas.* **2018**, *12*, 602. [CrossRef]

118. Stevenson, D.E.; Hurst, R.D. Polyphenolic phytochemicals—Just antioxidants or much more? *Cell Mol. Life Sci.* **2007**, *64*, 2900–2916. [CrossRef] [PubMed]

119. Amiot, M.J.; Riva, C.A.V. Effects of dietary polyphenols on metabolic syndrome features in humans: A systematic review. *Obes. Rev.* **2016**, *17*, 573–586. [CrossRef]

120. Hussain, T.; Tan, B.; Yin, Y.; Blachier, F.; Tossou, M.C.; Rahu, N. Oxidative Stress and Inflammation: What Polyphenols Can Do for Us? *Oxid. Med. Cell. Longev.* **2016**, *2016*. [CrossRef]

121. D'Archivio, M.; Filesi, C.; Varì, R.; Scazzocchio, B.; Masella, R. Bioavailability of the polyphenols: Status and controversies. *Int. J. Mol. Sci.* **2010**, *11*, 1321–1342. [CrossRef]

122. Rubió, L.; Macià, A.; Motilva, M.J. Impact of various factors on pharmacokinetics of bioactive polyphenols: An overview. *Curr. Drug Metab.* **2014**, *15*, 62–76. [CrossRef]

123. Zhang, H.; Yu, D.; Sun, J.; Liu, X.; Jiang, L.; Guo, H.; Ren, F. Interaction of plant phenols with food macronutrients: Characterisation and nutritional-physiological consequences. *Nutr. Res. Rev.* **2014**, *27*, 1–15. [CrossRef]

124. Manach, C.; Williamson, G.; Morand, C.; Scalbert, A.; Rémésy, C. Bioavailability and bioefficacy of polyphenols in humans. I. Review of 97 bioavailability studies. *Am. J. Clin. Nutr.* **2005**, *81* (Suppl. 1), 230S–242S. [CrossRef]

125. Salzanoa, S.; Checconia, P.; Hanschmannc, E.M.; Lillig, C.H.; Bowler, L.D.; Chan, P.; Vaudry, D.; Mengozzi, M.; Coppo, L.; Sacre, S.; et al. Linkage of inflammation and oxidative stress via release of glutathionylated peroxiredoxin-2, which acts as a danger signal. *Proc. Natl. Acad. Sci. USA* **2014**, *111*, 12157–12162. [CrossRef]

126. Kim, H.Y.; Nam, S.Y.; Hong, S.W.; Kim, M.J.; Jeong, H.J.; Kim, H.M. Protective effects of rutin through regulation of vascular endothelial growth factor in allergic rhinitis. *Am. J. Alle.* **2015**, *29*, 87–94. [CrossRef]

127. Chen, W.Y.; Huang, Y.C.; Yang, M.L.; Lee, C.Y.; Chen, C.J.; Yeh, C.H.; Pan, P.H.; Horng, C.T.; Kuo, W.H.; Kuan, Y.H. Protective effect of rutin on LPS-induced acute lung injury via down-regulation of MIP-2 expression and MMP-9 activation through inhibition of Akt phosphorylation. *Int. Immunopharmacol.* **2014**, *22*, 409–413. [CrossRef]

128. Qin, M.; Chen, W.; Cui, J.; Li, W.; Liu, D.; Zhang, W. Protective efficacy of inhaled quercetin for radiation pneumonitis. *Exp. Ther. Med.* **2017**, *14*, 5773–5778. [CrossRef] [PubMed]

129. Schreckinger, M.E.; Wang, J.; Yousef, G.; Lila, M.A.; Gonzalez de Mejia, E. Antioxidant capacity and in vitro inhibition of adipogenesis and inflammation by phenolic extracts of Vaccinium floribundum and *Aristotelia chilensis*. *J. Agric. Food Chem.* **2010**, *58*, 8966–8976. [CrossRef] [PubMed]

130. Paredes-López, O.; Cervantes-Ceja, M.L.; Vigna-Pérez, M.; Hernández-Pérez, T. Berries: Improving human health and healthy aging, and promoting quality life A review. *Plant Foods Hum. Nutr.* **2010**, *65*, 299–308. [CrossRef] [PubMed]

131. Cespedes, C.L.; Pavon, N.; Dominguez, M.; Alarcon, J.; Balbontin, C.; Kubo, I.; El-Hafidi, M.; Avila, J.G. The chilean superfruit black-berry *Aristotelia chilensis* (Elaeocarpaceae), Maqui as mediator in inflammation-associated disorders. *Food Chem. Toxicol.* **2017**, *108*, 438–450. [CrossRef]

132. Moon, H.D.; Kim, B.H. Inhibitory effects of Aristotelia chilensis water extract on 2,4-Dinitrochlorobenzene induced atopic-like dermatitis in BALB/c Mice. *Asian Pac. J. Allr. Immunol.* **2019**. [CrossRef]

133. Vergara, D.; Ávila, D.; Escobar, E.; Carrasco-Pozo, C.; Sánchez, A.; Gotteland, M. The intake of maqui (*Aristotelia chilensis*) berry extract normalizes H2O2 and IL-6 concentrations in exhaled breath condensate from healthy smokers—an explorative study. *Nutr. J.* **2015**, *14*, 27. [CrossRef] [PubMed]

134. Aguirre, M.C.; Delporte, C.; Backhouse, N.; Erazo, S.; Letelier, M.E.; Cassels, B.K.; Silva, X.; Alegría, S.; Negrete, R. Topical anti-inflammatory activity of 2a-hydroxy pentacyclic triterpene acids from the leaves of Ugni molinae. *Bioorg. Med. Chem.* **2006**, *14*, 5673–5677. [CrossRef]

135. Arancibia-Radich, J.; Peña-Cerda, M.; Jara, D.; Valenzuela-Bustamante, P.; Goity, L.; Valenzuela-Barra, G.; Silva, X.; Garrido, G.; Delporte, C.; Seguel, I. Comparative study of anti-inflammatory activity and qualitative-quantitative composition of triterpenoids from ten genotypes of *Ugni molinae*. *Bol. Latinoam. Caribe Plantas Med. Aromát.* **2016**, *15*, 274–287.

136. López, J.; Vega-Gálvez, A.; Rodríguez, A.; Uribe, E.; Bilbao-Sainz, C. Murta (*Ugni molinae* turcz.): A review on chemical composition, functional components and biological activities of leaves and fruits. *Chil. J. Agric. Anim. Sci.* **2018**, *34*, 1–14.

137. Rubilar, M.; Jara, C.; Poo, Y.; Acevedo, F.; Gutierrez, C.; Sineiro, J.; Shene, C. Extracts of maqui (*Aristotelia chilensis*) and murta (*Ugni molinae* Turcz.): Sources of antioxidant compounds and r-glucosidase/r-amylase inhibitors. *J. Agric. Food Chem.* **2011**, *59*, 1630–1637. [CrossRef] [PubMed]

138. Rojo, L.E.; Ribnicky, D.; Logendra, S.; Poulev, A.; Rojas-Silva, P.; Kuhn, P.; Dorn, R.; Grace, M.H.; Lila, M.A.; Raskin, I. In Vitro and in Vivo Anti-Diabetic Effects of Anthocyanins from Maqui Berry (*Aristotelia chilensis*). *Food Chem.* **2012**, *131*, 387–396. [CrossRef]

139. Schön, C.; Wacker, R.; Micka, A.; Steudle, J.; Lang, S.; Bonnländer, B. Bioavailability Study of Maqui Berry Extract in Healthy Subjects. *Nutrients* **2018**, *10*, 1720. [CrossRef] [PubMed]

140. Vegt, D.F.; Dekker, M.J.; Ruhe, G.H.; Stehouwer, D.C.; Nijpels, G.; Bouter, M.L.; Heine, J.R. Hyperglycemia is associated with all-cause and cardiovascular mortality in the Hoorn population: The Hoorn study. *Diabetologia* **1999**, *42*, 926–931. [CrossRef]

141. Durand, M.J.; Gutterman, D. Diversity in mechanisms of endothelium-dependent vasodilation in health and disease. *Microcirculation* **2013**, *20*, 239–247. [CrossRef] [PubMed]

142. Jofré, I.; Pezoa, C.; Cuevas, M.; Scheuermann, E.; Freires, I.A.; Rosalen, P.L.; de Alencar, S.M.; Romero, F. Antioxidant and Vasodilator Activity of *Ugni molinae* Turcz. (Murtilla) and Its Modulatory Mechanism in Hypotensive Response. *Oxid. Med. Cell. Longev.* **2016**, *2016*, 6513416. [CrossRef]

143. O'Keefe, J.H.; Bell, D.S. Postprandial hyperglycemia/hyperlipidemia (postprandial dysmetabolism) is a cardiovascular risk factor. *Am. J. Cardiol.* **2007**, *100*, 899–904. [CrossRef] [PubMed]

144. Cifuentes, F.; Palacios, J.; Paredes, A.; Nwokocha, C.R.; Paz, C. 8-Oxo-9-Dihydromakomakine Isolated from Aristotelia chilensis Induces Vasodilation in Rat Aorta: Role of the Extracellular Calcium Influx. *Molecules* **2018**, *23*, 3050. [CrossRef] [PubMed]

145. Davinelli, S.; Bertoglio, J.C.; Zarrelli, A.; Pina, R.; Scapagnini, G. A Randomized Clinical Trial Evaluating the Efficacy of an Anthocyanin-Maqui Berry Extract (Delphinol®) on Oxidative Stress Biomarkers. *J. Am. Coll. Nutr.* **2015**, *34* (Suppl. 1), 28–33. [CrossRef] [PubMed]

146. Cenativ. Available online: http://cenativ.utalca.cl/noticias/nuevas-variedades-de-maqui-abren-opcion-para-cultivo-del-fruto/ (accessed on 28 May 2019).

147. ODEPA—Oficina de Estudios y Políticas Agrarias. Boletín de Frutas y Hortalizas Procesadas. Available online: https://www.odepa.gob.cl/wp-content/uploads/2019/03/BFrutaprocesada0319.pdf (accessed on 13 May 2019).

148. Infor. Available online: http://www.infor.gob.cl/ (accessed on 26 May 2019).

149. Seguel, I.; Peñaloza, E.; Gaete, N. Colecta y caracterización molecular de germoplasma de murta (Ugni molinae Turcz.). en Chile. *Agro Sur* **2000**, *28*, 32–41. [CrossRef]

150. Alfaro, S.; Mutis, A.; Palma, R.; Quiroz, A.; Seguel, I.; Scheuermann, E. Influence of genotype and harvest year on polyphenol content and antioxidant activity in murtilla (*Ugni molinae* Turcz.) fruit. *J. Soil Sci. Plant Nutr.* **2013**, *13*, 67–78. [CrossRef]

151. EU. Available online: http://ec.europa.eu/food/safety/novel_food/catalogue/search/public/index.cfm# (accessed on 16 May 2019).

152. Slow Food Foundation for Biodiversity. Ark of Taste (Chile): "Purén White Strawberry". 2015. Available online: http://www.fondazioneslowfood.com/en/slow-food-presidia/puren-white-strawberry/ (accessed on 25 July 2015).

153. FIA. Available online: http://www.fia.cl/inia-y-fia-lanzan-polo-territorial-para-fortalecer-industria-de-colo rantes-naturales/ (accessed on 26 May 2019).

Article

Discovery of Unexpected Sphingolipids in Almonds and Pistachios with an Innovative Use of Triple Quadrupole Tandem Mass Spectrometry

Federico Maria Rubino [1], Michele Dei Cas [1], Monica Bignotto [1], Riccardo Ghidoni [1,2], Marcello Iriti [3] and Rita Paroni [1,*]

[1] Dipartimento di Scienze della Salute, Universita' degli Studi di Milano, I-20142 Milano, Italy; federico.rubino@unimi.it (F.M.R.); michele.deicas@unimi.it (M.D.C.); monica.bignotto@unimi.it (M.B.); riccardo.ghidoni@unimi.it (R.G.)
[2] Aldo Ravelli Center for Neurotechnology and Experimental Brain Therapeutics, Dipartimento di Scienze della Salute, Universita' degli Studi di Milano, I-20142 Milano, Italy
[3] Dipartimento di Scienze Agrarie e Ambientali—Produzione, Territorio, Agroenergia, Universita' degli Studi di Milano, I-20133 Milano, Italy; marcello.iriti@unimi.it
* Correspondence: rita.paroni@unimi.it

Received: 11 December 2019; Accepted: 19 January 2020; Published: 21 January 2020

Abstract: The densely packed storage of valuable nutrients (carbohydrates, lipids, proteins, micronutrients) in the endosperm of nuts and seeds makes the study of their complex composition a topic of great importance. Ceramides in the total lipid extract of some ground almonds and pistachios were searched with a systematic innovative discovery precursor ion scan in a triple quadrupole tandem mass spectrometry, where iso-energetic collision activated dissociation was performed. Five descriptors were used to search components with different C18 long chain bases containing different structural motifs (d18:0, d18:1, d18:2, t18:0, t18:1). The presence of hexoside unit was screened with a specific neutral loss experiment under iso-energetic collision activated dissociation conditions. The discovery scans highlighted the presence of two specific hexosyl-ceramides with a modified sphingosine component (d18:2) and C16:0 or C16:0 hydroxy-fatty acids. The hexosyl-ceramide with the non-hydroxylated fatty acid seemed specific of pistachios and was undetected in almonds. The fast and comprehensive mass spectrometric method used here can be useful to screen lipid extracts of several more seeds of nutraceutical interest, searching for unusual and/or specific sphingosides with chemically decorated long chain bases.

Keywords: ceramides; lipids; functional food; mass spectrometry; nutraceuticals; traditional food

---

1. Introduction

The densely packed content of carbohydrates, lipids, proteins, and other micronutrients in the endosperm of nuts and seeds is the storage of nutrients for the development and early growth of the seed into a new plantlet [1,2]. This characteristic makes them a complete and healthy food for animals and humans. Therefore, due to their traditional importance as a food staple in many countries and as a highly prized nutraceutical complement of health-promoting diets and of value-added traditional foods, the study of their composition is a topic of great importance [3]. Composition studies are especially important to relate the contents of micronutrients to health status [4,5] and to confirm the presence of specific components in complex preparations [6].

In particular, the ceramides present in the fatty component of nuts display a degree of chemical diversity in the sphingoid bases, fatty acid substituents, and C-1 appendages [7]. Limited information is available on occurrence and levels of ceramides in nuts. Miraliakbari and Shahidi et al. [8] reported that almond and pistachio oils contain 240 and 330 mg/100 g of total sphingolipids, respectively, measured utilising TLC-FID as a non-specific method of quantification. Using 750 Da as a mean representative value for sphingolipid molecular masses, this value corresponds to approximately 500 nmol/100 g. By using LC-MS/MS, we found 200 pmol/g = 20 nmol/100 g of ceramide species with the "standard" (i.e., mammalian) sphingoid base, 1,3-dihydroxy-D4-C18 sphingosine [9]. Even accounting for inefficient extraction and unspecific measurement, this discrepancy suggests that there is a major pool of sphingolipids unaccounted for in nuts.

The measured "standard" ceramides are not the only derivatives of sphingoid bases that are currently known to occur in nuts. Sang et al. [10] used high-resolution 1D and 2D NMR data to identify in almonds a monoglucocerebroside with a modified sphingoid base, sphinga-4,8-dienine. Fang et al. [6] employed liquid chromatography coupled to electrospray mass spectrometry with in-source fragmentation to identify further compounds in several plants, among which were almonds. Among those identified in almonds, there are ceramides and cerebrosides with trihydroxy bases with zero or one double bond, mainly 4-hydroxy-8-sphingenine (t18:1), amide-linked to very long chain fatty acids, with or without an α-hydroxy group. In this study, cerebrosides, expressed as d18:2-C16:0H-GLU, were measured at 68 µg/g (approximately 100 nanomoles/g, or 10 µmoles/100 g); thus, at a level that is at least two orders of magnitude higher than sphingosine d18:1 ceramides [6].

To systematically search for unexpected ceramides in lipid extracts, we propose an innovative use of the triple quadrupole tandem mass spectrometry based on a precursor ion scan with iso-energetic collision activated dissociation, aimed at systematic discovery and preliminary characterization of the main sphingosine components of almond and pistachio. Specifically, we searched for components with different C18 long chain bases containing different structural motifs (d18:0, d18:1, d18:2, t18:0, t18:1). The presence of hexoside unit was screened with a specific neutral loss experiment under iso-energetic collision activated dissociation conditions. Connectivity confirmation was achieved by fragment ions analysis with accurate mass measurements.

2. Results

2.1. Systematic Discovery of Sphingolipids in Almonds and Pistachios

We modified our standard LC-MS-MS method used for targeted ceramide measurement [9] to allow the systematic identification (discovery) of ceramides and ceramide hexosides with different sphingosine and fatty acid components. Modifications include:

(a) The use of full scan Precursor Ion (PI scan) of five O" ceramide reporter fragments (*m/z* 262, 264, 266, 280, 282; Table 1) to highlight ceramides with modified sphingosines;

(b) The use of a full scan Neutral Loss of $C_6H_{12}O_6$ hexose (180.2 Da) to ascertain whether the putatively highlighted ceramides have a hexose attached unit;

(c) The use of a collision energy ramp synchronized to the scan of Q1 in the PI and NL modes (iso-energetic Precursor Ion, *i*-PI, and iso-energetic Neutral Loss, *i*-NL) to analyze all ceramides at the same value of effective collision energy (see Appendix A);

(d) An extended isocratic step (in respect to the previously published conditions [9]) at full gradient strength during the UPLC analysis of sphingolipids (total analysis time 22 min) to investigate the possible presence of ceramides with much longer fatty acids.

Table 1. Structures of the screened long chain bases and of the corresponding reporter ion fragments.

ID	Core Structure of Long Chain Base (LCB)	LCB (MW) and O″ (m/z)
A	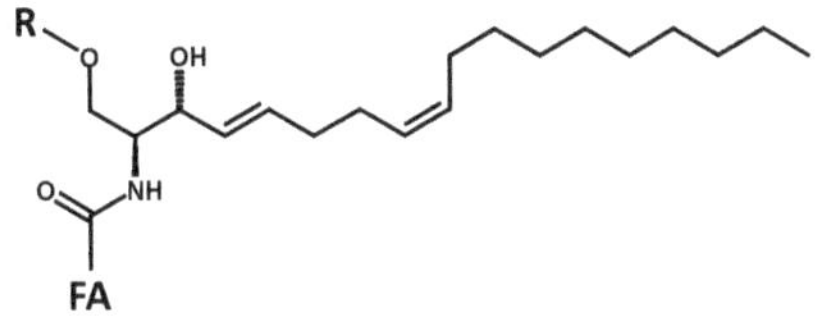	d18:2-4,8 $C_{18}H_{35}NO_2$ MW 297.49 O″ $C_{18}H_{32}N]^+$ m/z 262.22
B	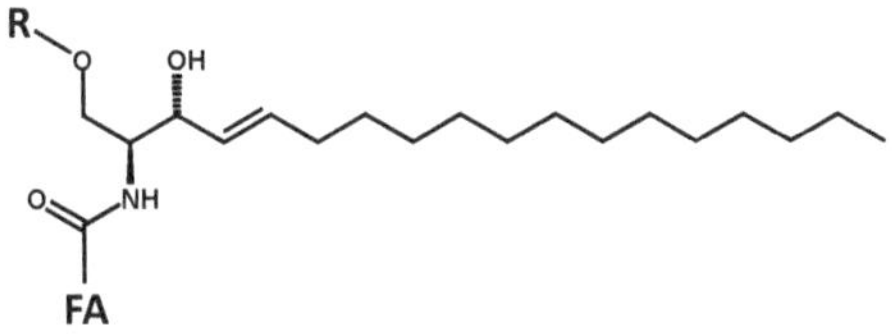	d18:1-4 $C_{18}H_{37}NO_2$ MW 299.50 O″ $C_{18}H_{34}N]^+$ m/z 264.22
C	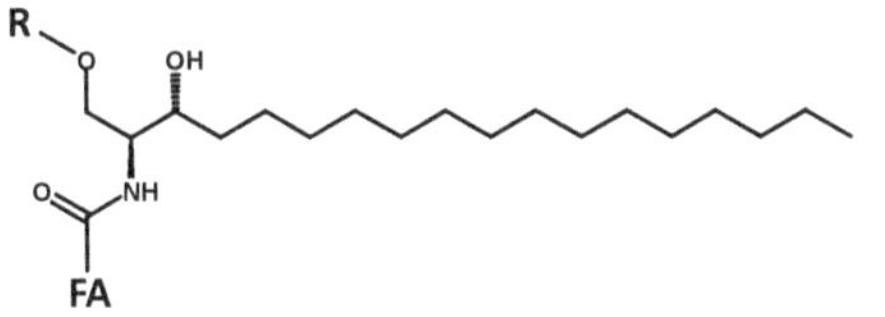	d18:0 $C_{18}H_{39}NO_2$ MW 301.52 O″ $C_{18}H_{36}N]^+$ m/z 266.22
D	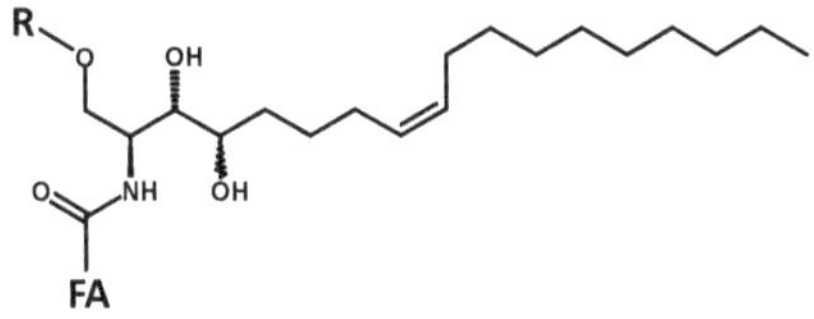	t18:1-8 $C_{18}H_{37}NO_3$ MW 315.50 O″ $C_{18}H_{34}NO]^+$ m/z 280.22
E	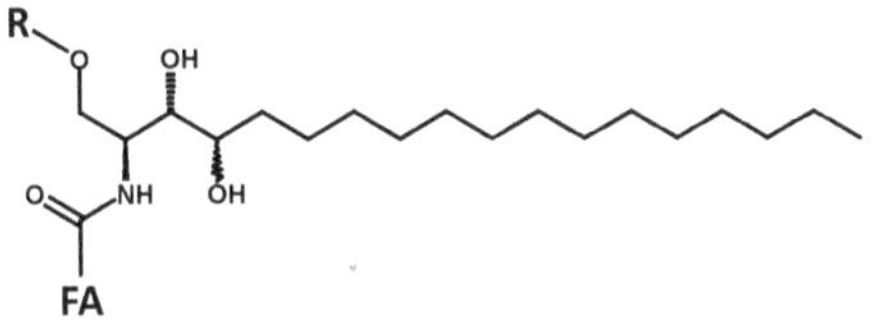	t18:0 $C_{18}H_{39}NO_3$ MW 317.52 O″ $C_{18}H_{36}NO]^+$ m/z 282.22

R = H (ceramide); R = Hex (cerebroside); FA = fatty acid residue

Fragment O″ = [LCB − 2 H_2O] × H^+

The systematic search for ceramides and hexose-linked ceramides that contain each of the five hypothesized LCBs was performed, for each sample, in four separate chromatographic runs. Each run alternates two 1-s *i*-PI scans of m/z 264 (reporter ion of d18:1 internal standards and of ceramides with d18:1 sphingosine) and one for each of the other LCBs (m/z 262, 266, 280, 282). This procedure allowed for an estimation of the amount of unexpected ceramides and carbohydrate-linked ceramides with reference to those of the Cer12:0 and Cer12:0-Glc that show in the *i*-PI scan of m/z 264.

To verify whether any identified compound is really a ceramide hexoside (there is no possibility to ascertain hexose stereochemistry and configuration with this set of experiments), an iso-energetic Neutral Loss scan (*i*-NL) of the hexose fragment ($C_6H_{12}O_6$, MW 180.2 Da) was paired to the Precursor Ion scan that yielded the putative ceramide hexoside.

Figure 1 shows the results of this set of experiments applied to a few interesting samples used to illustrate the discovery procedure. The other samples showed a similar pattern, with only differences in the relative intensity of the chromatographic peaks.

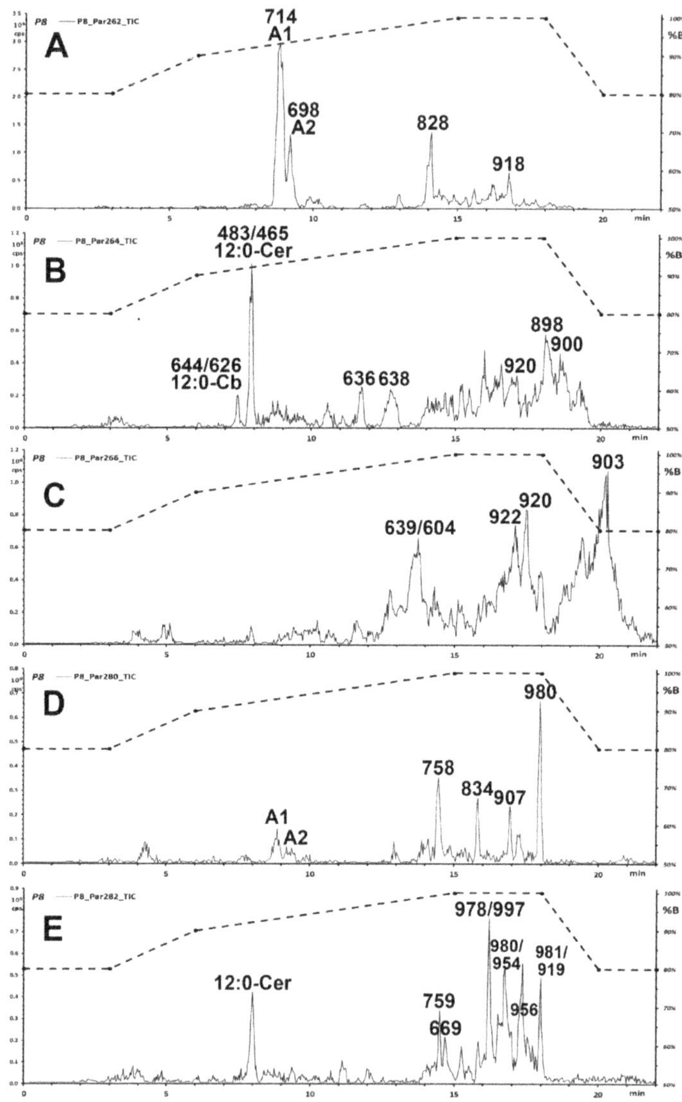

Figure 1. Chromatographic traces of the iso-energetic Precursor Ion scans of five long chain bases ((**A–E**), see Table 1 for their structures) in the extract of a pistachio cultivar. The dashed trace displays the chromatographic gradient, as volume fraction of acetonitrile (%B; right vertical axis). Numbers on top of prominent peaks are the *m/z* of the most intense ion signals. Labelled are the added internal standards (12:0-Cb and 12:0-Cer) and the newly identified hexosyl-ceramides A1 and A2.

Figure 2 illustrates the corresponding integrated mass spectra of each iso-energetic Precursor Ion scan of Figure 1. Main signals that may correspond to likely unexpected ceramide species awaiting identification are highlighted.

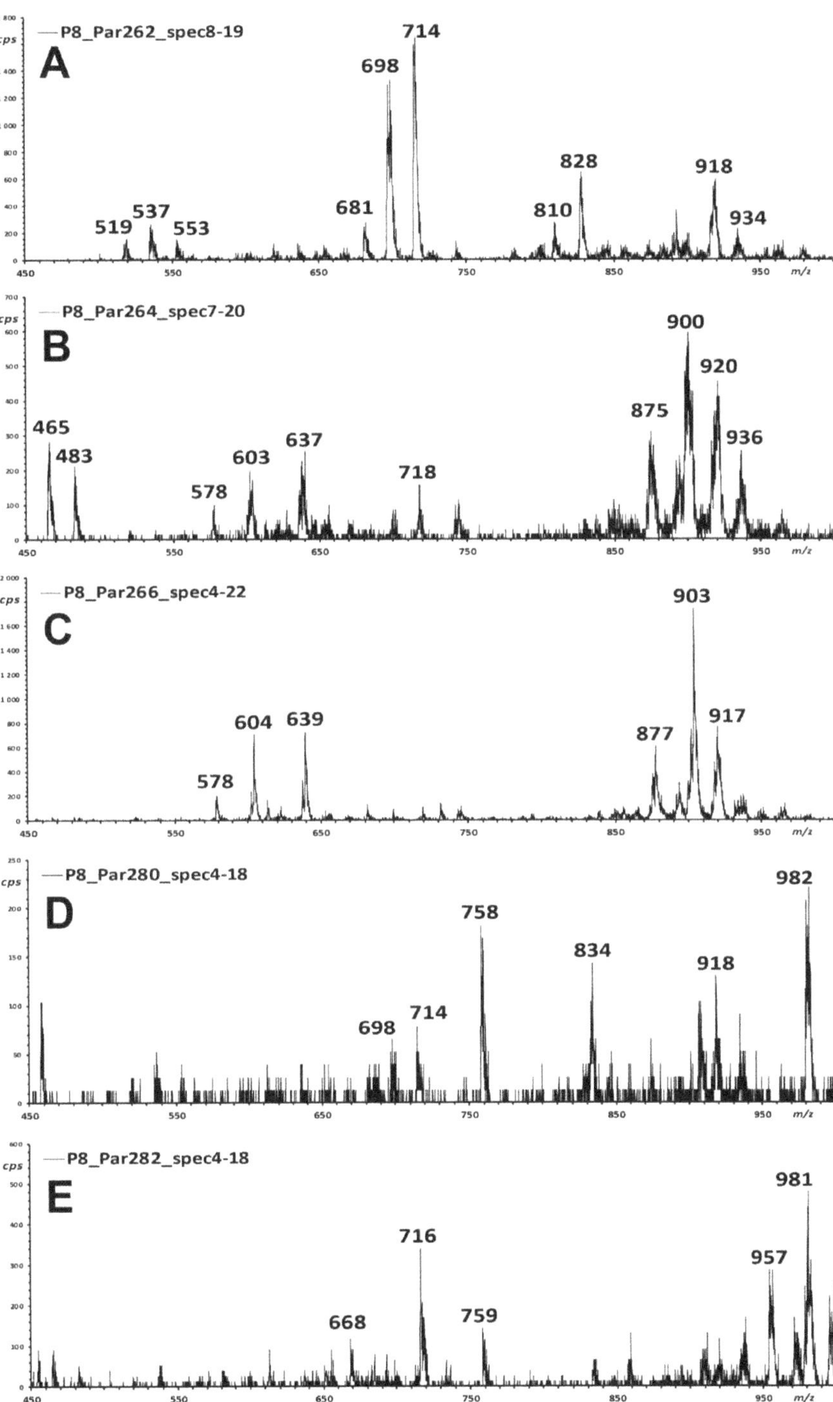

Figure 2. Integrated Precursor Ion mass spectra of the iso-energetic scan five long chain bases ((**A–E**), see Table 1 for their structures) in the extract of a pistachio cultivar. Integration is performed, for all experiments, within the 3.5 to 21 min interval of the chromatographic traces of Figure 1. Labelled are the *m/z* values of some main signals.

The measurement of more than 50 chromatographic runs for the 15 examined samples highlighted the presence of at least two main chemical species. A major, earlier-eluting one (A1, retention time, RT ≈ 8.5 min; relative retention time, RRT = 1.18 vs. 12:0-Cb; MH$^+$ 714) is present in all pistachio and almond samples (Figure 1A is a representative pistachio sample). A minor, later-eluting one (A2, RT ≈ 9.1 min; RRT = 1.23 vs. 12:0-Cb; MH$^+$ 698) is present only in pistachios, but not in almonds. The two compounds featured prominently in the Par262 trace (Figure 1A) and yielded weaker peaks in the Par280 (Figure 1D) trace.

According to the general fragmentation pattern of protonated ceramides reported in Figure 3 [11], the glycosidic nature of compounds A1 and A2 is warranted by the coelution also in the NL180 experiment (Figure 4A,C) and by the observation of the same precursor ions in the NL180 and Par262 spectra (Figure 4B,D).

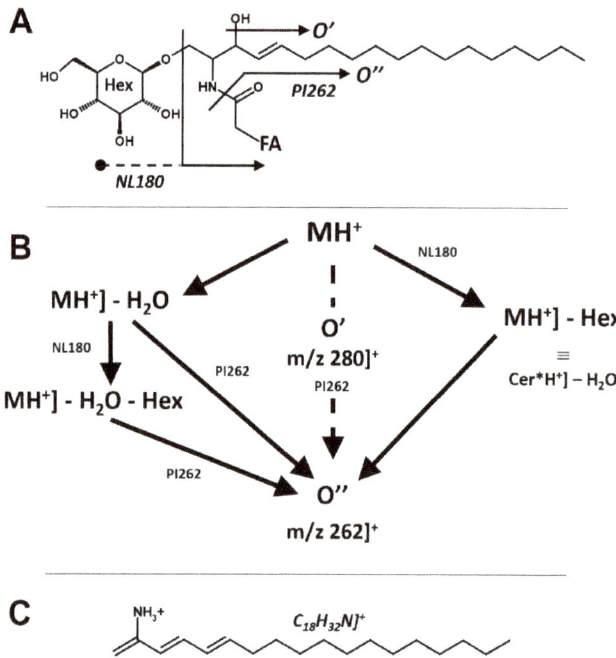

Figure 3. Overall fragmentation pattern of protonated hexosyl-ceramides, exemplified for those with the "standard" 1,3-dihydroxy-C-18:1 long chain base. Panel (**A**): correspondence of fragments to motifs in the molecular structure. Panel (**B**): development of decomposition processes and triple quadrupole scan experiments used to highlight individual molecules from their fragmentation. Panel (**C**): formal connectivity of the ceramide pivot fragment O''.

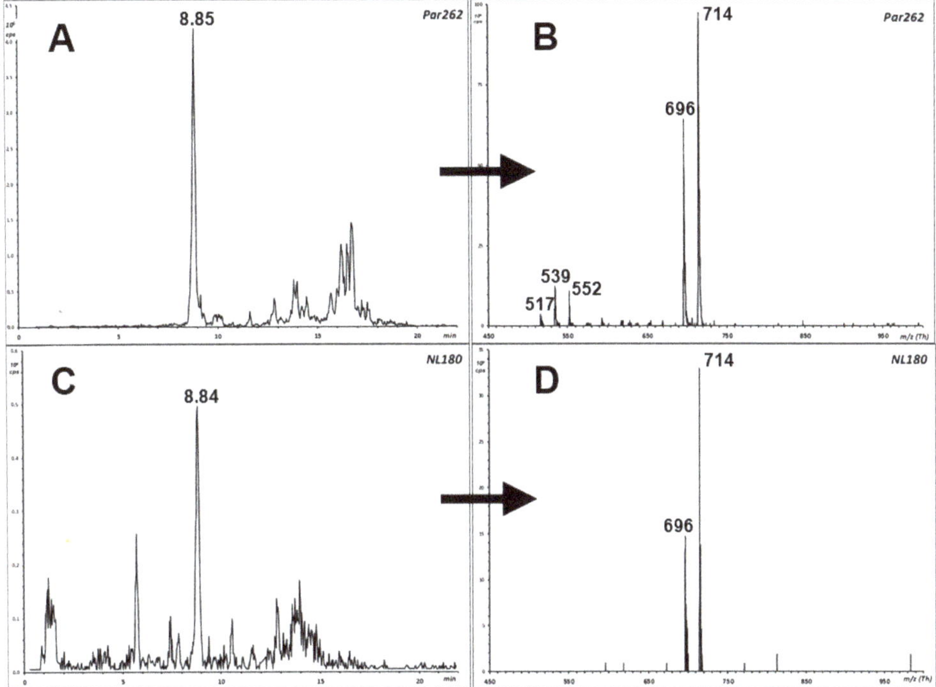

Figure 4. Confirmation of an unexpected hexose-containing ceramide species in the extract of an almond cultivar. The highlighted peak in panels (**A**,**C**) is that of compound A1. *m/z* 714 is the molecular mass (MH$^+$) in both experiments (Par262, panel (**B**), and NL180, panel (**D**). *m/z* 696 derives from MH$^+$ by water loss (panels (**B**,**D**)). In Par262 (panel (**B**)) *m/z* 552 and 534 derive from the previous ones by the loss of hexose.

Their connectivity was investigated by recording their Fragment Ion spectra in the EPI mode (Q3 used as a LIT) in separate chromatographic runs (A1, A2, Figure 5).

According to the fragmentation pattern summarized in Figure 3, compounds A1 and A2 thus consist of:

- A modified sphingosine that carries a further unsaturation in the C18 chain (d18:2, fragment ion at *m/z* 262 Th);
- The C-1 of sphingosine is linked to a hexose (paired losses of 180 Da from MH$^+$ and [MH − H$_2$O]$^+$ ions);
- The 2-amino group of the sphingosine is linked to two saturated C16 fatty acids (mass difference between *m/z* 262 and the fragments generated by hexose loss), one of which (earlier-eluting compound A1) carries an additional hydroxyl group.

The occurrence of a linked hexose unit at C-1 was confirmed by performing an NL-180 scan concurrent to the Par262 scan, observing peaks in both traces at the retention times of the two compounds and MH$^+$ and [MH − H$_2$O]$^+$ ions (Figure 3).

Final evidence of the identity of the precursor and fragment ions detected in the triple quadrupole instrument was achieved by re-analyzing two almond and two pistachio extracts under comparable chromatographic conditions in a quadrupole-ToF instrument that measures *m/z* at a resolution close to 30,000. Figure S1 and Table S1 summarize the results. High-resolution measurements match the elemental composition of the examined precursor ions and of the expected fragments within 2–26 ppm, thus confirming the interpretation of the LIT fragment ion spectra of Figure 4 and the connectivity of the two compounds.

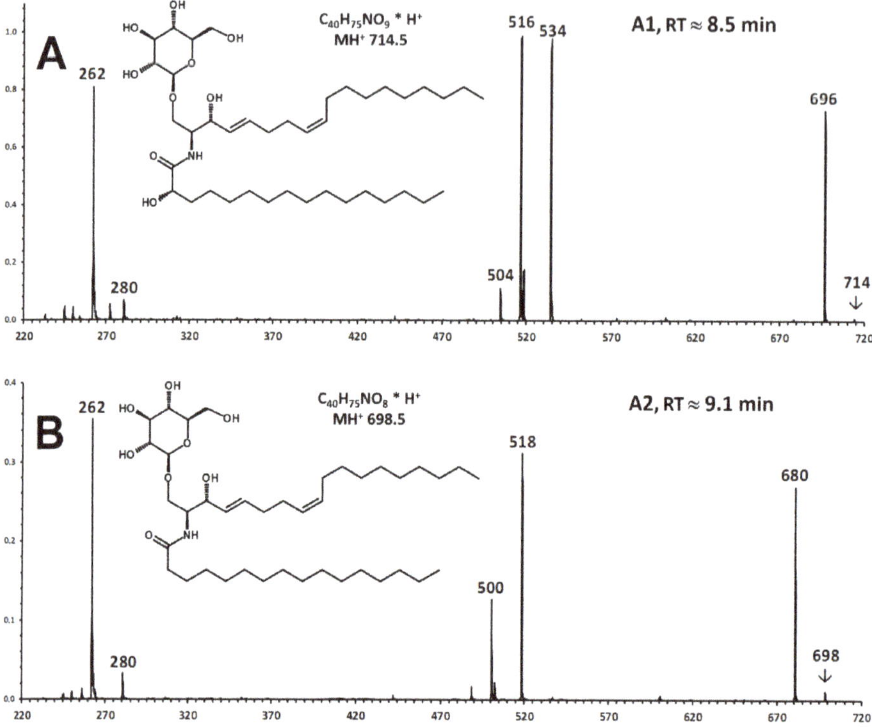

Figure 5. Enhanced Product Ion (EPI, CE 30 DV) spectra of unexpected ceramide species in the extract of a pistachio cultivar. (**A**) Compound A1 is that eluting at 8.5 min in the chromatographic trace A of Figure 1; (**B**) compound A2 is that eluting at 9.1 min.

Previous researchers already extracted and identified similar compounds in almonds [12]. The authors assigned the connectivity of the additional long chain base double bond as z-Δ8 and identified the stereochemistry of the appended hexose with the extensive use of Nuclear Magnetic Resonance experiments.

2.2. Levels of the Discovered Sphingolipids in Almonds and Pistachios

The analysis of three almond and nine pistachio samples highlighted that A2 (which contains a saturated C16 fatty acid) occurs in both almonds and pistachios, while A1 (which contains a hydroxylated C16:0-h fatty acid) is specific of pistachio. Their abundance in the sample could be estimated in the absence of authentic purified reference ceramides, by comparing the peaks areas to that of C12:0 cerebroside (peak at 7.3 min in panel B of Figure 1) added as an internal standard. In pistachios, components A1 and A2 were estimated on the average at 19.3 and 5.2 µg/g of extracts, respectively. In almonds, compound A1 is estimated on the average at 24.2 µg/g of extracts. The levels of the two characteristic ceramides in the individual nut samples are reported in Table 2.

Table 2. Levels of two specific ceramides in three almond and nine pistachio samples.

Sample ID	A1	A2
M1	13.65	n.d.
M2	22.48	n.d.
M3	36.47	n.d.
P1	39.97	13.64
P2	17.37	2.87
P3	14.69	4.53
P4	14.35	3.63
P5	15.95	4.04
P6	17.26	4.09
P7	19.40	3.74
P8	22.24	7.13
P9	12.52	2.85

A1: RT 8.5 min; MH$^+$ 715; A2: RT 9.1 min; MH$^+$ 699; Results in µg/g (expressed as Cer12:0-Glc equivalents); n.d.: not detected.

While Reisberg and collaborators [12] did not perform an analytical assessment, but rather isolated and characterized component A1, they could purify approximately 120 mg from 14 kg of almonds. This amount corresponds to 8.6 µg/g and is close to that measured in our samples.

2.3. Interference of Triglycerides in the Discovery of Ceramides in Almonds and Pistachios

As apparent from the chromatographic traces and integrated mass spectra of Figures 1 and 2, an intense signal was detected for materials that eluted later than 15 min and that contained arrays of signals up to the *m/z* 1000 limit of the Q1 quadrupole scan. The time limit of 15 min is the elution time of the ceramide with the longest (C24:0) fatty acid of the standard series. Grounding on the specificity of the even-*m/z* O″ reporter fragments for nitrogen-containing molecules, this region of the chromatograms was explored to detect possible sphingoid compounds with more complex structures.

In particular, the Par264 trace (Figure 2B) featured even-*m/z* molecular signals with a molecular mass around 900 Da (*m/z* 898, 900, and 902). Possible connectivity at this molecular size that is compatible with a ceramide structure might belong to epidermosides, a class of very large ceramides with a ω-hydroxylated fatty acid, to which a further fatty acid is connected through an ester bond. Such compounds are found in skin, as components of the lipid mixture of the stratum corneum [13]. However, the Fragment Ion spectra recorded for the main *m/z* signals did not match the expected behaviour for the anticipated structures. In particular, the observed fragmentation did not yield the fragment pair spaced by 18 Da that derives from the two fragmentation modes of the long chain ester bond. In addition, the compounds eluted earlier than expected for their molecular size based on the behaviour of the standard series of saturated and unsaturated sphingosine ceramides [14].

Since the occurrence of these likely spurious signals may bias the discovery of high-mass, unexpected sphingosine derivatives, a brief investigation was undertaken to understand their origin. In particular, the mass spectral features of the cluster at 18–20 min elution time in the Par264 experiment (Figures 1B and 2B) matched the loss of the elements of fatty acid-ammonium ion pairs in the Product ion experiment (diglyceride fragments at *m/z* 599.5, 601.6, 603.5, data not shown), which is the well-documented behaviour of triglycerides [15]. High-resolution measurement in the Q-ToF mass spectrometer on the same extract confirmed the postulated identity of the observed signals as ammonium ion adducts of triglycerides (Figure S2). In particular, TG ammonium adducts at *m/z* 900.8 (TG 54:4) were mainly composed of dioleoyl-linoleoyl-glycerol (OOL). This identification is confirmed by the occurrence of the expected losses of fatty acids in the tandem mass spectra, with the generation of diglyceride fragments at *m/z* 601 and 603 in a 3:1 ratio (Figure S2). Under the adopted collision conditions, the acylium fragments do not form.

To investigate whether co-extracted triglycerides may be the source of the high-mass signals erroneously identified as putative complex ceramide (Figures 1B and 2B), a source scan at baseline unit resolution in Enhanced Mass Scan (EMS) mode was recorded (data not shown). Most m/z signals recorded at the retention times of the considered chromatographic peaks correspond to a cluster of eluting triglycerides (TG) with ^{12}C isotopomers at m/z 896.6, 898.7, and 900.5. The observed ion clusters corresponded to ammoniated triglycerides with a total of 54 fatty acid carbon atoms and a number of unsaturations that range from six to four, or three, two, and one units of linoleic acid (m/z 896.6 TG 54:6, m/z 898.7 TG 54:5, m/z 900.5 TG 54:4). The precursor ions detected in the Par264 scan (Figures 1B and 2B) at m/z 898.5, 900.5, and 902.6 correspond to the [M + 2] isotopomers of each ammoniated triglyceride (TG 54:6, TG 54:5, TG 54:4; Figure S2C,E), which mainly contain two ^{13}C carbon atoms in the molecule. Therefore, the observed signal (Figures 1B and 2B), that interferes with the O″ reporter fragment ion of sphingosine ceramides, may correspond to the ^{13}C$_1$-isotopomer (m/z 264) of the acylium fragment of linoleic acid all-^{12}C $C_{18}H_{31}O^+$ (m/z 263) that is present as a minor process in the spectra of triglycerides ionized as ammonium-adducted species [16].

To test whether this hypothesis holds, the same sample was analyzed with a simultaneous recording of the Precursor Ion spectra of m/z 263 (the acylium fragment of linoleic acid, all-^{12}C $C_{18}H_{31}O^+$) and of m/z 264 (^{13}C$_1$-isotopomer). As apparent in the two superimposed graphs of Figure S3, the chromatographic profiles of m/z 263 and 264 closely match in the time frame later than approximately 9 min, and their intensities are in an approximately 4:1 ratio. Only in the time frame between 7.5 and 8.5 min, where the abundant 12:0 ceramides elute, the m/z 264 signal (the O″ fragment of S18d:1) predominates over the background signal and yields the expected molecular precursors of the S18d:1 species. In the time frame later than approximately 9 min, the two Precursor Ion spectra record signals of ammoniated triglyceride and diglyceride from the bulk of the un-fractionated lipid extract (spectra not shown). In particular, between 18 and 20 min, the Par263 scan (Figure S3) detects domes of molecular signals, among which those at m/z 896.9, 898.8, and 900.9 (TG 54:6, TG 54:5 and TG 54:4). The corresponding Par264 scan detects domes of molecular signals (m/z 897–902, spectra not shown) with an intensity that is approximately 25% of that of the corresponding signals of the Par263 scan. This intensity ratio closely corresponds to that expected for the occurrence of the ^{13}C isotopomers of the fatty acid in the fragment signals of the precursor triglycerides (Figure S4).

3. Discussion

Research on the characterization of oils and solid fats (butter) derived from nuts is extensive [17], due to the long traditional use of these products and of increasing consumption as alternatives to dairy sources. Most analyses focus on the composition of major components, such as the presence of fatty acids [18,19]), of triglycerides and of some minor components with nutraceutical properties, such as vitamins and antioxidants [8,20,21].

Sphingolipid research in plants has highlighted the presence of modified long chain bases, mainly t18:1 and d18:2, linked to commonly occurring and more rare odd-carbon and hydroxylated fatty acids [6,10,12,22,23].

Observation of ceramides with a modified long chain base and with specific fatty acids may increase the variety of chemical indicators that can be employed to confirm the presence of specific high-value nuts, such as pistachios and almonds, in food preparations, and to highlight contamination or adulteration with extraneous materials [2,24,25]. Several techniques are currently used for this task, including direct Near-Infra-Red [26] and Raman [27] spectroscopy, mineral element pattern [28], and chemometric evaluation of untargeted LC-HRMS analysis [29].

It is of note that selective scans in the Precursor Ion and Neutral Loss modes are unique of the tandem triple quadrupole (and, earlier, of some magnetic) mass spectrometer configuration. The sole limitation of the use of triple quadrupole is the unit resolution of the mass filters. As highlighted in our case, spurious signals can be generated from the unexpected contribution of isotopic components of bulk constituents. This approach allowed observing and understanding for the first time the interference of a high load of triglycerides in the selective discovery search for trace level nitrogen-containing lipophilic secondary metabolites by triple quadrupole tandem mass spectrometry methods. In addition, the comparatively lower sensitivity of the iso-energetic Precursor Ion approach is inherent to the use of a scan mode, rather than MRM, and is an unavoidable trade-off of the demonstrated higher coverage of different chemical species [30,31].

Complete characterization of the discovered ceramide hexosides entails recognition of the hexose stereochemistry and confirmation of the position and stereochemistry of the structural modifications of the fatty acid and of the sphingosine. These structural details cannot be easily differentiated by simple, *on the fly* recording of tandem mass spectra but need isolation of more substantial amounts of purified materials by mass spectrometry guided preparative methods.

4. Materials and Methods

4.1. Reagents, Chemicals, and Standards

Sphingolipid standards, including ceramides with d18:1D4 sphingosine long-chain base (LCB) and saturated/singly unsaturated straight-chain even-carbon fatty acids (12–24 FA) and C12:0 cerebroside were purchased from Avanti Polar Lipids (Alabaster, AL, USA). Methanol, ethanol, acetonitrile, ammonium formate, and formic acid (all analytical grade) were supplied from Merck (Darmstadt, Germany). Water was MilliQ-grade (Millipore, Milford, MA, USA).

4.2. Plant Material

Three selected almond (A6, A7, A8) and nine pistachio (P7–P15) extracts were analyzed in this study. Their codes and their main characteristics are reported in Table 3.

Table 3. Pistachio and almond samples examined in this study. The first column refers to the blinded analysis code and the second column to the sample code.

ID	Sample	Nut Product	Origin	Characteristics
			Pistachio (*Pistacia vera* L.)	
P1	P7	Shelled	Bronte DOP (Sicily, Italy)	Not roasted, not salted
P2	P8	Shell	USA	Roasted, salted
P3	P9	Shell	Non EU	Organic, roasted, not salted
P4	P10	Shell	Iran	Not roasted, not salted
P5	P11	Shell	USA	Roasted, not salted
P6	P12	Shelled	Noberasco	Not roasted, not salted
P7	P13	Flour	Iran	
P8	P14	Flour	Italy	
P9	P15	Flour	Italy	
ID	**Sample**	**Nut Product**	**Origin**	**Characteristics**
			Almond (*Prunus dulcis* Mill.)	
M1	A6	Shelled	California (USA)	Died
M2	A7	Shelled	California (USA)	Dried
M3	A8	Shelled	California (USA)	Dried

4.3. Sphingolipid Extraction Procedure

The total lipid extracts were prepared as described elsewhere [9]. Briefly, after the addition of 10 μL of IS (Cer C12, gluCer C12 20 mM), the powder (250 mg) was extracted in a microtube with an O-ring seal screw with 500 μL of methanol, 100 μL water, and 250 μL chloroform [32]. The samples were sonicated for 30 min and incubated overnight in an oscillator bath at 48 °C. The supernatant was evaporated under a stream of nitrogen. The residues were dissolved in 150 μL of methanol and then centrifuged for 10 min at 13,000 rpm. The clean supernatant (130 μL) was transferred into the autosampler vials, and 10 μL were directly injected in LC-MS/MS.

4.4. LC-MS/MS Instrumentation

Two separate sets of mass spectrometry measurements were performed in different computerized integrated systems, each consisting of a liquid chromatograph coupled to a tandem mass spectrometer.

Discovery of ceramides and ceramide hexosides with a diverse long chain base was accomplished in a hybrid triple quadrupole-linear ion trap tandem mass spectrometer. The integrated liquid chromatograph system was an UltiMate® 3000 LC Systems (Dionex™, Sunnyvale, CA, USA), with an autosampler, binary pump, and column oven (Thermo Fisher Scientific, Waltham, MA, USA). The tandem mass spectrometer was an AB Sciex 3200 QTRAP LC-MS/MS instrument with electrospray ionization (ESI) TurboIonSpray™ source (AB Sciex Framingham, MA, USA). Instruments were managed with manufacturers' software (Analyst software (version 1.6.2) and according to manufacturers' instructions. Results exported as .txt files were post-processed in custom spreadsheets, as needed. The principle of the innovative scan modes employed in this instrument is briefly described in Appendix A.

A set of confirmative measurements used a Shimadzu UPLC interfaced to a TripleTOF 6600 (Sciex, MA, USA) high-resolution hybrid quadrupole time-of-flight mass spectrometer equipped with Turbo Spray IonDrive, under essentially the same chromatographic conditions (2.3, *v. infra*). The resolution was close to 30,000 according to the manufacturer's software evaluation. The main instrument parameters were: CUR 35, GS1 55, GS2 35, capillary voltage 5.5 kV, source temperature 350 °C, declustering potential (DP) 50 eV. Two analysis modes were applied. One ("*targeted*" approach) recorded, within each 1-s measurement cycle, the source spectrum (*m/z* 200–1400; 250 ms accumulation time) and the fragment ion spectra of selected precursors only (100 ms accumulation time). The other ("*untargeted*" approach) recorded, within each 1-s measurement cycle, the source spectrum, and the fragment ion spectra of the ten most abundant precursor ions that occur in the previous source spectrum. Collision energy was set at 30 V on a nitrogen gas target. Data were extracted with the proprietary PeakView data management software and exported to spreadsheets for further evaluation.

4.5. Separation and Detection of Sphingolipids by LC-MS/MS

Separation of the lipid extract containing ceramides was accomplished in an ACQUITY UPLC BEH C-8 Column, 130 Å, 1.7 μm, 2.1 mm × 100 mm (Waters, Millford, MA, USA) preceded by a security guard cartridge. The flow rate was 0.3 mL/min; the autosampler and the column oven were kept at 15 °C and 30 °C, respectively, the operating pressure was 450 Psi.

The two mobile phases were: phase A, 2 mM ammonium formate in water and phase B 1 mM ammonium formate in methanol, both containing 0.2% formic acid (*v/v*). A multi-linear extended gradient with a total analysis time of 22 min was programmed: the column was equilibrated with 80% (B), increased to 90% (B) in 3 min, held for 3 min, increased to 99% (B) in 9 min, held for 3 min, back to the initial conditions in 2 min, and kept for 2 min at 80% (B).

Mass spectrometry was performed in the positive ion mode (ESI+). The ion spray voltage was set at 5.5 kV, and the source temperature was set at 300 °C. Nitrogen was used as a nebulizing gas (GS 1, 45 psi), turbo spray gas (GS 2, 50 psi), and curtain gas (25 psi). Source spectra were recorded in separate experiments in the Enhanced MS (EMS) mode at a scan speed of 1000 Da/s that yielded baseline separation of unit mass peaks in the *m/z* range 450–1000. The collision-activated dissociation (CAD) MS-MS experiment used nitrogen as collision gas at the low pressure setting (1.2×10^{-5} Torr).

4.6. Untargeted Discovery LC-MS/MS Analysis by Iso-Energetic Precursor Ion and Neutral Loss Scan in a Triple Quadrupole

An untargeted discovery method was developed to investigate the presence of unexpected sphingolipids in the samples. Briefly, the method employs Precursor Ion (PI) scans of the analysing quadrupole (Q1), while holding the selecting quadrupole (Q3) to transmit the reporter fragment ions generated by CID of protonated ceramides with different long chain bases (fragment O").

Those selected are at *m/z* 264.4 (d18:1D4 sphingosine, [11]) and isomeric d18:1D8 [6], *m/z* 262.4 (d18:2D4,8); *m/z* 280.4 (t18:1D8, [6]).

Another set of experiments employed Neutral Loss scans, whereby both Q1 and Q3 mass filters are simultaneously scanned at the same rate, with a fixed offset of transmitted *m/z*, which corresponded to the mass of a specific molecular unit. In this case, the neutral fragment was the $C_6H_{12}O_6$ unit (MW 180.2) of hexose.

The upper scan range of Q1 was initially *m/z* 450–700 in 1 s (measurement of the standard ceramide panel), and later, the upper *m/z* value was raised to *m/z* 1000 to allow analysis of ceramides with as many 60 carbon atoms. During each 1-s scan of Q1 (or of linked Q1 and Q3), the CAD potential (q2-Q1) was synchronously ramped in order that, at each *m/z* value of the precursor ion transmitted by Q1, the effective collision energy was held constant (iso-energetic, or *i*-CID). The particular selected value of the effective collision energy was selected following a spectroscopic study of model ceramides. Actual employed values are of 1.6 eV for the Precursor Ion scans (generation of O" fragment from MH^+) and 1.0 eV for the Neutral Loss scan (loss of the hexose unit from the MH^+). Conversion of centre-of-mass collision energy to instrument voltage was accomplished with a re-arranged form of the standard equation (a more detailed explanation is reported in Appendix A). Typical ranges of collision voltage for the Precursor Ion scans were 27.3–44.5 DV for the low mass range (*m/z* 450–750) and 27.3–55.9 DV for the extended mass range (*m/z* 450–1000). The corresponding values for the Neutral Loss scan were 27.3–44.5 DV for the low mass range (*m/z* 450–750) and 27.3–55.9 DV for the extended mass range (*m/z* 450–1000).

4.7. Relationship of Molecular Structure to Chromatographic Retention

To assist and to confirm the identification of unexpected ceramide species, a descriptor of the relationship of molecular structure to chromatographic retention was established. A standard mixture of ceramides ranging from C 12:0 to 24:0 (2.6 μM each) was injected under described conditions. Natural logarithm of capacity factor *k'*, defined as the number of column void volumes necessary to elute each compound, was plotted against the natural logarithm of both the total number of carbon atoms in the ceramide and of the sole number of carbons of the FA (graph reported in Figure S5).

The void volume was estimated from the geometric parameters of the column and from the mobile phase flow. A correction factor for fractional column packing was identified from technical literature, as 0.51 of the physical column volume. Relative retention times (RRT) were calculated with reference to the elution of C12 glucosyl ceramide (cerebroside, 12:0-Cb for short).

4.8. Isotope Pattern Calculation

Isotope pattern calculation was accomplished with an online freeware calculator (https://www.envipat.eawag.ch/index.php; last accession 4 November 2019). The software allows modulation of the mass resolution to envision the profile of the isotope cluster in the employed instrumental conditions. In this instance, the resolution was modulated at two different values, according to the employed mass spectrometer. A value of 500 (DM/M), slightly lower than that effective in the triple quadrupole instrument, but sufficient to match the measured ion profiles, was employed to simulate the isotopic envelopes for the first tier of measurements. A value of 10,000 (DM/M) was employed to calculate the abundances and accurate *m/z* of precursor and fragment ions anticipated for the identified compounds, and to compare with high-resolution measurements.

5. Conclusions

This report furthers our previous one on the targeted analysis of ceramides (d18:0, d18:1) in almond and pistachio [9]. This technique exploits the performance of the triple quadrupole tandem mass spectrometer well beyond its traditional use by analytical chemists.

Success with this approach will prompt a complete investigation of the structural details of the observed phytochemicals that cannot be resolved with the analytical and spectroscopic techniques employed in this preliminary survey. Semi-preparative chromatographic separation of crude lipid preparations driven by the selective scan will enable to obtain enriched fractions of the unexpected components. The use of high-energy collision induced dissociation will characterize the connectivity of the ceramide and fatty acid subunits [11], and deglycosidation will enable to separately identify the linked hexose.

Availability of a fast and comprehensive method to screen lipid extracts for trace sphingosides with chemically decorated long chain bases will be applied to analyze several more seeds of a nutraceutical interest that are the base of typical foods of the Italian lore.

Supplementary Materials: The following are available online at http://www.mdpi.com/2304-8158/9/2/110/s1. Figure S1: High-resolution mass spectrometry analysis for the confirmation of the unexpected ceramide species in the extract of a pistachio cultivar. Table S1: High-resolution measurement of fragment ion spectra of the four main ceramide species. Figure S2: High-resolution molecular (a, c, e) and fragment ion spectra (b, d, f) of triglycerides. Figure S3: Chromatographic traces of the Par263 and Par264 scans for a representative extract. Figure S4: Formation of protonated diglyceride and acylium fatty acid fragments from triglycerides. Figure S5: Plot of the relationship of chromatographic retention vs. molecular size for ceramides.

Author Contributions: Conceptualization, F.M.R.; methodology, F.M.R., M.D.C.; investigation, F.M.R., M.D.C., M.B.; resources, M.I., R.P.; data curation, F.M.R., M.D.C., M.B.; writing—original draft preparation, F.M.R., M.D.C, R.P.; writing—review and editing, F.M.R., M.D.C, R.P, R.G, M.I.; visualization, F.M.R., M.D.C.; supervision, R.P.; project administration, R.P., M.I.; funding acquisition, R.P., M.I., R.G. All authors have read and agreed to the published version of the manuscript.

Funding: This research received no external funding.

Acknowledgments: MDC receives support from the PhD program in Molecular and Translational Medicine of the Università degli Studi di Milano, Italy. Part of this work was carried out in OMICs, an advanced mass spectrometry platform established by the Università degli Studi di Milano.

Conflicts of Interest: The authors declare no conflict of interest.

Appendix A

Re-arranged form of the standard equation for energy transfer in Collision Induced Dissociation.

In Collision Induced Dissociation of ions, the relation of centre-of-mass, or "effective" (E_{CM}) to laboratory-frame or "nominal" (E_{lab}) collision energy depends:

- On the mass of the resting target gas in the collision cell (m_{TAR}) and
- On the mass (m/z) of the impinging (singly-charged) precursor ion (m_{PAR}),

according to the standard Equation (1).

$$E_{CM} = E_{lab} \times m_{TAR}/(m_{TAR} + m_{PAR})$$

For a fixed collision gas employed (e.g., m_{TAR}, nitrogen, MW 28, or argon, MW 40), and for each specific value of "laboratory frame" collision energy, the "centre-of-mass" collision energy decreases as the m/z value of the precursor ions increase. The variation is not linear, either with the m/z of the precursor ion, or with the value of the "laboratory frame" collision energy.

Standard algebraic passages allow re-formulating Equation (1) to the linear working Equation (2):

$$E_{lab} = E_{CM} + E_{CM} \times m_{PAR}/m_{TAR}.$$

This linear equation allows to "back"-calculate the necessary laboratory frame collision voltage at each value of m/z of the precursor ion corresponding to a specific constant value of centre-of-mass collision energy.

References

1. Hawker, J.S.; Buttrose, M.S. Development of the Almond Nut (*Prunus dulcis* (Mill.) DA Webb). Anatomy and Chemical Composition of Fruit Parts from Anthesis to Maturity. *Ann. Bot.* **1980**, *46*, 313–321. [CrossRef]
2. Summo, C.; Palasciano, M.; De Angelis, D.; Paradiso, V.M.; Caponio, F.; Pasqualone, A. Evaluation of the chemical and nutritional characteristics of almonds (*Prunus dulcis* (Mill). DA Webb) as influenced by harvest time and cultivar. *J. Sci. Food Agric.* **2018**, *98*, 5647–5655. [CrossRef]
3. Alasalvar, C.; Bolling, B.W. Review of nut phytochemicals, fat-soluble bioactives, antioxidant components and health effects. *Br. J. Nutr.* **2015**, *113*, S68–S78. [CrossRef]
4. Wang, D.D.; Toledo, E.; Hruby, A.; Rosner, B.A.; Willett, W.C.; Sun, Q.; Razquin, C.; Zheng, Y.; Ruiz-Canela, M.; Guasch-Ferré, M.; et al. Plasma ceramides, mediterranean diet, and incident cardiovascular disease in the PREDIMED trial (prevención con dieta mediterránea). *Circulation* **2017**, *135*, 2028–2040. [CrossRef] [PubMed]
5. Neeland, I.J.; Singh, S.; McGuire, D.K.; Vega, G.L.; Roddy, T.; Reilly, D.F.; Castro-Perez, J.; Kozlitina, J.; Scherer, P.E. Relation of plasma ceramides to visceral adiposity, insulin resistance and the development of type 2 diabetes mellitus: The Dallas Heart Study. *Diabetologia* **2018**, *61*, 2570–2579. [CrossRef] [PubMed]
6. Fang, F.; Ho, C.T.; Sang, S.; Rosen, R.T. Determination of sphingolipids in nuts and seeds by a single quadrupole liquid chromatography-mass spectrometry method. *J. Food Lipids* **2005**, *12*, 327–343. [CrossRef]
7. Michaelson, L.V.; Napier, J.A.; Molino, D.; Faure, J.D. Plant sphingolipids: Their importance in cellular organization and adaption. *Biochim. Biophys. Acta* **2016**, *1861*, 1329–1335.
8. Miraliakbari, H.; Shahidi, F. Antioxidant activity of minor components of tree nut oils. *Food Chem.* **2008**, *111*, 421–427. [CrossRef]
9. Paroni, R.; Dei Cas, M.; Rizzo, J.; Ghidoni, R.; Montagna, M.T.; Rubino, F.M.; Iriti, M. Bioactive phytochemicals of tree nuts. Determination of the melatonin and sphingolipid content in almonds and pistachios. *J. Food Compos. Anal.* **2019**, *82*, 103227. [CrossRef]
10. Sang, S.; Kikuzaki, H.; Lapsley, K.; Rosen, R.T.; Nakatani, N.; Ho, C.T. Sphingolipid and other constituents from almond nuts (*Prunus amygdalus* Batsch). *J. Agric. Food Chem.* **2002**, *50*, 4709–4712. [CrossRef]
11. Rubino, F.M.; Zecca, L.; Sonnino, S. Characterization of a Complex Mixture of Ceramides by Fast-Atom-Bombardment and Precursor and Fragment Analysis Tandem Mass-Spectrometry. *Biol. Mass Spectrom.* **1994**, *23*, 82–90. [CrossRef]
12. Reisberg, M.; Arnold, N.; Porzel, A.; Neubert, R.H.H.; Dräger, B. Production of Rare Phyto-Ceramides from Abundant Food Plant Residues. *J. Agric. Food Chem.* **2017**, *65*, 1507–1517. [CrossRef] [PubMed]
13. Motta, S.; Sesana, S.; Ghidoni, R.; Monti, M. Content of the different lipid classes in psoriatic scale. *Arch. Dermatol. Res.* **1995**, *287*, 691–694. [CrossRef]
14. Gaudin, K.; Chaminade, P.; Baillet, A. Structure-retention diagrams of ceramides established for their identification. *J. Chromatogr. A* **2002**, *973*, 69–83. [CrossRef]
15. Murphy, R.C.; James, P.F.; McAnoy, A.M.; Krank, J.; Duchoslav, E.; Barkley, R.M. Detection of the abundance of diacylglycerol and triacylglycerol molecular species in cells using neutral loss mass spectrometry. *Anal. Biochem.* **2007**, *366*, 59–70. [CrossRef] [PubMed]
16. Goodacre, R.; Vaidyanathan, S.; Bianchi, G.; Kell, D.B. Metabolic profiling using direct infusion electrospray ionisation mass spectrometry for the characterisation of olive oils. *Analyst* **2002**, *127*, 1457–1462. [CrossRef] [PubMed]
17. Gorrepati, K.; Balasubramanian, S.; Chandra, P. Plant based butters. *J. Food Sci. Technol.* **2015**, *52*, 3965–3976. [CrossRef]
18. Yildirim, A.N.; Akinci-Yildirim, F.; Şan, B.; Sesli, Y. Total Oil Content and Fatty Acid Profile of some Almond (Amygdalus communis L.) Cultivars. *Pol. J. Food Nutr. Sci.* **2016**, *66*, 173–178. [CrossRef]
19. Moayedi, A.; Rezaei, K.; Moini, S.; Keshavarz, B. Chemical compositions of oils from several wild almond species. *J. Am. Oil Chem. Soc.* **2011**, *88*, 503–508. [CrossRef]
20. Fernandes, G.D.; Gómez-Coca, R.B.; Pérez-Camino, M.D.C.; Moreda, W.; Barrera-Arellano, D. Chemical Characterization of Major and Minor Compounds of Nut Oils: Almond, Hazelnut, and Pecan Nut. *J. Chem.* **2017**, *2017*, 11. [CrossRef]
21. Evoli, L.D.; Lucarini, M.; Gabrielli, P.; Aguzzi, A.; Lombardi-boccia, G. Nutritional Value of Italian Pistachios from Bronte (Pistacia vera, L.), Their Nutrients, Bioactive Compounds and Antioxidant Activity. *Food Nutr. Sci.* **2015**, *6*, 1267.

22. Alasalvar, C.; Shahidi, F. Tree Nuts: Composition, Phytochemicals, and Health Effects: An Overview. In *Tree Nuts: Composition, Phytochemicals, and Health Effects*; CRC Press: Boca Raton, FL, USA, 2008; ISBN 978-1-4200-1939-1.

23. Valsecchi, M.; Mauri, L.; Casellato, R.; Ciampa, M.G.; Rizza, L.; Bonina, A.; Bonina, F.; Sonnino, S. Ceramides as possible nutraceutical compounds: Characterization of the ceramides of the moro blood orange (Citrus sinensis). *J. Agric. Food Chem.* **2012**, *60*, 10103–10110. [CrossRef] [PubMed]

24. Liang, L.; Wu, X.; Zhu, M.; Zhao, W.; Li, F.; Zou, Y.; Yang, L. Chemical composition, nutritional value, and antioxidant activities of eight mulberry cultivars from China. *Pharmacogn. Mag.* **2012**, *8*, 215–224. [CrossRef] [PubMed]

25. Esteki, M.; Ahmadi, P.; Vander Heyden, Y.; Simal-Gandara, J. Fatty acids-based quality index to differentiate worldwide commercial pistachio cultivars. *Molecules* **2019**, *24*, 58. [CrossRef] [PubMed]

26. Bernardi, N.; Benetti, G.; Campolongo, G.; Ferrari, G.; Palermo, R.; Vranic, B. Preliminary Assessment of Vegetable Oil Adulteration of Pistachio Paste by near Infrared Spectroscopy. *NIR News* **2014**, *25*, 20–21. [CrossRef]

27. Eksi-Kocak, H.; Mentes-Yilmaz, O.; Boyaci, I.H. Detection of green pea adulteration in pistachio nut granules by using Raman hyperspectral imaging. *Eur. Food Res. Technol.* **2016**, *242*, 271–277. [CrossRef]

28. Sezer, B.; Apaydin, H.; Bilge, G.; Boyaci, I.H. Detection of Pistacia vera adulteration by using laser induced breakdown spectroscopy. *J. Sci. Food Agric.* **2018**, *99*, 2236–2242. [CrossRef]

29. Çavuş, F.; Us, M.F.; Güzelsoy, N.A. Assesing Pistachio Nut (Pistacia vera L.) Adulteration with Green Pea (Pisum sativum L.) by Untargeted Liquid Chromatography-(Quadrupole-Time of Flight)-Mass Spectrometry Method and Chemometrics. Available online: https://dergipark.org.tr/en/pub/bursagida/issue/40169/477784 (accessed on 20 January 2020).

30. Canela, N.; Herrero, P.; Mariné, S.; Nadal, P.; Ras, M.R.; Rodríguez, M.Á.; Arola, L. Analytical methods in sphingolipidomics: Quantitative and profiling approaches in food analysis. *J. Chromatogr. A* **2016**, *1428*, 16–38. [CrossRef]

31. Markham, J.E.; Jaworski, J.G. Rapid measurement of sphingolipids from Arabidopsis thaliana by reversed-phase high-performance liquid chromatography coupled to electrospray ionization tandem mass spectrometry. *Rapid Commun. Mass Spectrom.* **2007**, *21*, 1304–1314. [CrossRef]

32. Dalmau, N.; Jaumot, J.; Tauler, R.; Bedia, C. Epithelial-to-mesenchymal transition involves triacylglycerol accumulation in DU145 prostate cancer cells. *Mol. Biosyst.* **2015**, *11*, 3397–3406. [CrossRef]

Sample Availability: Small samples of some almond and pistachio specimens are available from the authors upon reasonable request.

Article

Wild Italian *Prunus spinosa* L. Fruit Exerts In Vitro Antimicrobial Activity and Protects Against In Vitro and In Vivo Oxidative Stress

Luisa Pozzo [1,*], Rossella Russo [1], Stefania Frassinetti [1], Francesco Vizzarri [2,3], Július Árvay [4], Andrea Vornoli [1,5], Donato Casamassima [2], Marisa Palazzo [2], Clara Maria Della Croce [1] and Vincenzo Longo [1]

[1]   Institute of Agricultural Biology and Biotechnology (IBBA), National Research Council, 56124 Pisa, Italy; rossella.wilbur@gmail.com (R.R.); stefania.frassinetti@ibba.cnr.it (S.F.); vornolia@ramazzini.it (A.V.); clara.dellacroce@ibba.cnr.it (C.M.D.C.); v.longo@ibba.cnr.it (V.L.)
[2]   Department of Agricultural, Environmental and Food Sciences, University of Molise, 86100 Campobasso, Italy; francesco.vizzarri@uniba.it (F.V.); casamassima.d@unimol.it (D.C.); m.palazzo@unimol.it (M.P.)
[3]   Department of Agricultural and Environmental Sciences, University of Bari Aldo Moro, 70126 Bari, Italy
[4]   Department of Chemistry, Faculty of Biotechnology and Food Sciences, Slovak University of Agriculture, 94976 Nitra, Slovakia; julius.arvay@uniag.sk
[5]   Cesare Maltoni Cancer Research Center, Ramazzini Institute, Bentivoglio, 40138 Bologna, Italy
*   Correspondence: luisa.pozzo@ibba.cnr.it; Tel.: +39-0503152702

Received: 8 November 2019; Accepted: 17 December 2019; Published: 19 December 2019

Abstract: Polyphenol-rich foods could have a pivotal function in the prevention of oxidative stress-based pathologies and antibacterial action. The purpose of this study was to investigate the in vitro antimicrobial activity, as well as the in vitro and In Vivo antioxidant capacities of wild *Prunus spinosa* L. fruit (PSF) from the southeast regions of Italy. The total phenolic content (TPC) was quantified, and the single polyphenols were analyzed by HPLC-DAD, showing high rutin and 4-hydroxybenzoic acid levels, followed by gallic and trans-sinapic acids. PSF extract demonstrated antimicrobial activity against some potentially pathogenic Gram-negative and Gram-positive bacteria. Besides, we investigated the cellular antioxidant activity (CAA) and the hemolysis inhibition of PSF extract on human erythrocytes, evidencing both a good antioxidant power and a marked hemolysis inhibition. Furthermore, an In Vivo experiment with oxidative stress-induced rats treated with a high-fat diet (HFD) and a low dose of streptozotocin (STZ) demonstrated that PSF has a dose-dependent antioxidant capacity both in liver and in brain. In conclusion, the wild Italian *Prunus spinosa* L. fruit could be considered a potentially useful material for both nutraceutical and food industries because of its antioxidant and antimicrobial effects.

Keywords: wild Italian *Prunus spinosa* L. fruit; blackthorn; phenolic compounds; antimicrobial; antioxidant

---

1. Introduction

Considering that traditional foods are increasingly believed healthy and wholesome, food manufacturers are developing new food products returning to natural products and traditional recipes that will be attractive to the widest potential consumers [1,2]

Blackthorn (*Prunus spinosa* L.), which belongs to the *Rosaceae* family, is a perennial plant originally growing in temperate continental climate of the northern hemisphere, particularly widespread in the Mediterranean countries and in the southeast regions of Italy. It is used for treatment of many diseases due to its diuretic, spasmolytic, antimicrobial, and antioxidant activities [3]. Moreover, *Prunus spinosa* L. fruit (PSF) is used for the production of various traditional jams and beverages such as juice, wine, tea, and distillates in food industry [4]. It contains substantial quantities of phenolic antioxidants, including, in particular, flavonols, phenolic acids, and coumarin derivatives [5].

Epidemiological investigations demonstrated that diets rich in plant polyphenols protect against diabetes, osteoporosis, cardiovascular, and neurodegenerative diseases [6]. Dietary compounds and specific polyphenol-rich foods could have a pivotal function in the prevention of diseases associated with oxidative stress by increasing the circulation of antioxidant compounds and neutralizing the reactive oxygen species, due to their number and position of hydroxyl groups [7]. Protein nitration, lipid peroxidation, chronic inflammation, and oxidative damage to DNA may be prevented by polyphenols, which results in vasodilatory, vasoprotective, anti-atherogenic, antithrombotic, and anti-apoptotic effects, as free radical scavengers, metal chelators, inhibitors of pro-inflammatory enzymes, and modifiers of cell signaling pathways [8].

Recently, new alternatives have become desirable, and plants metabolites have been screened for antimicrobial agents for treatment of infectious diseases, due to the development of antibiotic resistance by pathogenic bacteria [9]. For instance, in several studies, dietary polyphenols have been reported to exert an antibacterial activity [10].

In the present study, the potential biological activities of wild blackthorn fruit from southeast regions of Italy were investigated. Considering the PSF as a potential natural source of phenolic compounds, this work was designed to study its in vitro antimicrobial, antioxidant, and antihemolytic activities. Moreover, for the first time, the In Vivo protective effect and antioxidant capacity of PSF in high-fat diet (HFD) and streptozotocin (STZ)-induced oxidative stress has been studied.

2. Materials and Methods

Blackthorn fresh fruits were obtained from wild orchards of the Campobasso (Italy) area in October 2015. Taxonomic identification of plant material was confirmed by Prof. Elisabetta Brugiapaglia from Department of Agricultural, Environmental and Food Sciences, University of Molise, Campobasso, Italy.

2.1. Chemicals and Reagents

All solvents and chemicals were of analytical grade. Nutrient Broth (NB), Nutrient Agar (NA), Mueller Hinton Broth (MHB), Mueller Hinton Agar (MHA), McFarland standard were purchased from Oxoid (Basingstone, UK). 6-hydroxy-2,5,7,8-tetramethylchroman-2-carboxylic acid (Trolox), 2,2′-azobis (2-amidinopropane) dihydrochloride (AAPH), 2,7-dichlorodihydrofluorescein diacetate (DCFH-DA), dinitrophenylhydrazine (DNPH), trichloroacetic acid (TCA), perchloric acid (PCA), thiobarbituric acid (TBA), 1,1,3,3-tetramethoxypropane (TEP), guanidine hydrochloride, ortho-phthalaldehyde (OPA), reduced glutathione (GSH), phosphoric acid, potassium dihydrogen phosphate (KH_2PO_4), hydrochloric acid (HCl), streptozotocin (STZ), ethanol, ethyl acetate, and methanol from Sigma-Aldrich (St. Louis, MO, USA). All HPLC analytical standards, including protocatechuic acid, syringic acid, rutin, ellagic acid, cynaroside, daidzein, neochlorogenic acid, chlorogenic acid, vitexin, trans *p*-coumaric acid, trans-sinapic acid, trans ferulic acid, rosmarinic acid, resveratrol, apigenin, myricetin, quercetin, and kaempferol, were bought from Sigma-Aldrich (St. Louis, MO, USA). Phosphate buffer saline (PBS) was bought from VWR (Radnor, PA, USA).

2.2. Plant Material Preparation

After the pits were removed, the frozen fruits were lyophilized and crushed in a mortar, and the powder was stored at −20 °C. Briefly, 1 g of sample was extracted with 10 mL of water for 2 h on a horizontal shaker Unimax 2010 (Heidolph Instruments, GmbH, Schwabach, Germany) for in vitro antioxidant activity, polyphenols quantification, and antimicrobial activity. PSF extracts were centrifuged (2300× g at 4 °C for 20 min) (Jouan CR3i centrifuge, Newport Pagnell, UK), and the supernatants were collected.

For the In Vivo experiment, the lyophilized and powdered fruit was dissolved in water.

2.3. Total Phenolic Content (TPC) and Polyphenols Quantification by HPLC-DAD

For the determination of total phenolic content in water extract, we followed the Singleton and colleagues protocol [11]. The concentration of polyphenols was expressed as mg of gallic acid equivalents (GAE)/g of dry weight (d.w.).

Prior to HPLC analysis, the extracts were filtered through syringe filters Q-Max (0.22 μm, 25 mm, PVDF) (Frisenette ApS, Knebel, Denmark) into the HPLC vials. The HPLC apparatus consisted of an Agilent 1260 Infinity HPLC (Agilent Technologies GmbH, Waldbronn, Germany) quaternary solvent manager coupled with degasser (G1311B), sampler manager (G1329B), Diode Array Detector (G1315C), column manager (G1316A). The analytical column was a Waters Cortecs endcapped RP-C18 column (150 mm × 4.6 mm × 2.7 μm particle size; Waters Corp., Milford, MA, USA). The analyses were carried out at 30 °C by a gradient system with a mobile phase of 0.1% ortho-phosphoric acid in deionised water (C) and acetonitrile gradient grade (D) at a flow rate of 0.60 mL/min, and the injection volume was 5 μL. The gradient elution was as follows: 0–1 min (90% C and 10% D), 1–5 min (85% C and 15% D), 5–10 min. (80% C and 20% D), 10–12 min. (80% C and 20% D), 12–20 min (30% C and 70% D), and 20–25 min (30% C and 70% D). The post-run was set at 3 min. The samples were kept at 4 °C in the sampler manager. The detection wavelengths were set at 265 nm (gallic acid, 4-hydroxibenzoic acid, rutin, and genistein), 320 nm (chlorogenic acid, caffeic acid, trans-*p*-coumaric acid, trans-sinapic acid, trans-ferulic acid, rosmarinic acid, resveratrol), and 372 nm (myricetin, quercetin and kaempferol). Data were analyzed by Agilent Open Lab Chem Station software for LC 3D systems.

2.4. HPLC-DAD Method Validation

Reference phenolic compounds were dissolved in HPLC purity methanol and diluted to appropriate concentration ranges (5–50 μg/mL). The linearity of each calibration curve was assessed by linear regression analysis. The limit of detection (LOD) and quantification (LOQ) were estimated by measuring signal-to-noise ratio of the individual peak of each standard compound. The LOD and LOQ were calculated according to the International Conference on Harmonisation guidelines [12].

2.5. Antimicrobial Activity

2.5.1. Growth Conditions of Pathogenic Bacteria

The bacterial strains were supplied by the American Type Culture Collection (ATCC). The antimicrobial activity of PSF extract was studied on three Gram-negative bacteria, specifically *Escherichia coli* (ATCC 25922), *Salmonella enterica* ser. *typhimurium* (ATCC 14028), and *Enterobacter aerogenes* (ATCC 13048), and two Gram-positive bacteria, *Enterococcus faecalis* (ATCC 29212) and *Staphylococcus aureus* (ATCC 25923).

2.5.2. Antimicrobial Activity

The growth inhibition of selected bacteria exerted by PSF extract was determined according to Delgado Adámez and colleagues [13], with some modifications.

The tested bacteria were cultured in MHB at 37 °C for 16 h and diluted to match the turbidity of 0.5 McFarland standard. Fifty microliters of bacterial suspensions (about 1–5×10^5 CFU/mL) was added to 100 μL of MHB and to 100 μL of blackthorn extract (0, 0.25, 0.50, 0.75, and 1.00 mg/mL) in a 96-well plate. A negative control was included on each microplate. A positive control of bacterial growth inhibition consisting of two antibiotics, vancomycin (10 μg/mL) for Gram-positive and gentamicin (10 μg/mL) for Gram-negative bacteria was added to the microplate. The plates were incubated at 37 °C for 24 h. Afterwards, the optical density (OD) at 600 nm was determined by a microplate reader (Eti-System fast reader Sorin Biomedica, Modena, Italy). The percentage of growth inhibition was calculated as follows:

$$\% \text{ growth inhibition} = 100 - (OD_S/OD_C) \times 100 \tag{1}$$

where OD_S is the optical density of the sample and OD_C is the optical density of the negative control (PSF 0 mg/mL).

2.6. In Vitro Antioxidant Activity in Red Blood Cells (CAA-RBC) and Hemolysis Test

According to the regulations of "Fondazione G. Monasterio CNR-Regione Toscana", human blood samples were obtained from three healthy volunteers in ethylenediaminetetraacetic acid (EDTA)-treated tubes and centrifuged (2300× g at 4 °C for 10 min). Plasma and buffy coat were removed, and erythrocytes were washed twice with PBS pH 7.4.

The antioxidant activity of PSF extract (100 mg/mL) was evaluated in an in vitro system with a modified assay in red blood cells as described by Frassinetti and colleagues [14]. Each value was express as CAA units, as follows [12]:

$$\text{CAA unit} = 100 - (\int \text{SA} \backslash \int \text{CA}) \times 100 \qquad (2)$$

where $\int$SA is the integrated area of the sample curve and $\int$CA is the integrated area of the control curve.

Hemolysis of PSF extract (100 mg/mL) was analyzed according to the protocol described by Frassinetti and colleagues [15] using AAPH, a generator of peroxyl radicals, to cause the red blood cell lysis. The values reported are the percentage of hemolysis compared with the control.

2.7. Animal Study

2.7.1. In Vivo experiment

Male Wistar rats (200–230 g b.w.) were maintained with ad libitum access to food and drinking water for a 12 h light/dark cycle in cages at room temperature with the 55% relative humidity. Rats were divided into two groups: the control (CTR) group (n = 5), fed a standard diet (64% carbohydrates, 19% proteins, 7% minerals and vitamins, 6% fibers, and 4% fats; the fats percentage corresponds to the 11% of the diet-derived energy) and the high-fat diet (HFD) group, fed a high fat/cholesterol diet (48.7% carbohydrates, 28% fats, including 2% cholesterol, 13.8% proteins, 4.4% fibers, 5.1% minerals and vitamins, the fats percentage corresponds to the 55% of diet-derived energy). After 5 weeks, the animals of HFD group were treated with a single i.p. injection of streptozotocin (40 mg/kg) [16]. Twenty four rats, resulted to be diabetic with a plasma glucose concentration >250 mg/dL, continued to be fed a HFD diet for a further 4 weeks and were randomly divided into three groups: HFD (n = 8) group, PSF400 (n = 8) group, and PSF800 (n = 8) group (Figure 1). Rats from CTR and HFD groups were intragastrically administered the same volume of water; rats from PSF400 and PSF800 groups were intragastrically administered lyophilized PSF at different doses (400 mg/kg b.w. and 800 mg/kg b.w., respectively). The weight gain was calculated by initial and final weights. The rats were sacrificed and blood samples were collected by cardiac puncture under general anesthesia. Liver and brain tissues were stored at −80 °C. Hepatic lipids were quantified and oxidative stress markers were analyzed in liver and brain. Blood was centrifuged (2300× g for 15 min) to obtain serum samples for laboratory analysis. Local Ethical Committee approved all animal procedures in accordance with the European Communities Council Directive of 24 November, 1986 (86/609/EEC).

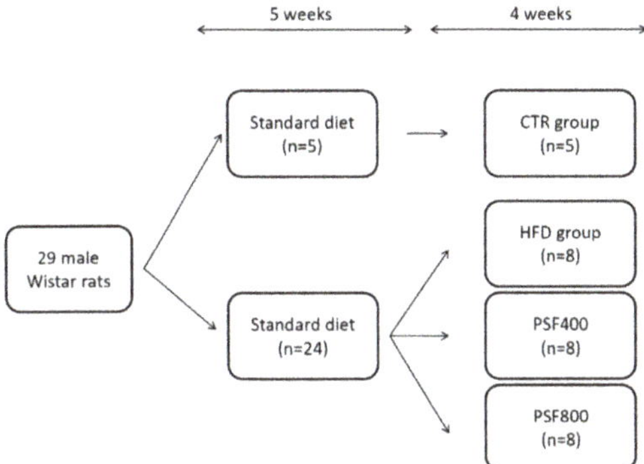

Figure 1. Study design of the In Vivo experiment. CTR, control group; HFD, high-fat diet group; PSF400, PSF800.

2.7.2. Biochemical Analysis

Serum analyses were performed by a semi-automatic analyzer for clinical chemistry (model ARCO, Biotecnica Instruments SPA, Rome, Italy) for aspartate aminotransferase (AST), alanine aminotransferase (ALT), total cholesterol, and triglycerides. Glucose levels were measured with a glucose meter (Accu-Chek® Roche, Mannheim, Germany), and insulin using a Rat Insulin ELISA kit (Mercodia, Uppsala, Sweden).

2.7.3. Hepatic Lipids Quantification

Fat content of the liver samples was determined by Folch and colleagues protocol [17], slightly modified. Liver samples from rats were homogenized with equal volumes of water and methanol. The resulting homogenate was subjected to three subsequent extractions in chloroform, followed by two washes with KCl 1 M and water. After complete evaporation and prolonged drying of the chloroform, fat content was weighed and expressed as mg/g tissue.

2.7.4. Oxidative Stress

Malondialdehyde (MDA) concentration of liver and brain samples was analyzed according to Seljeskog and colleagues [18], with some adaptations. An aliquot of 100 μL of homogenate sample was mixed with 0.1125 N PCA (300 μL) and 40 mM TBA (300 μL) for 10 sec and placed in a boiling water bath for 60 min. Methanol (600 μL) and 20% TCA (*w/v*) (200 μL) were added to the suspension and mixed for 10 sec, after cooling in a freezer at −20 °C for 20 min. The MDA content was quantified in the supernatant (7000× g for 6 min) by fluorimeter (Perkin Elmer LS-45, Perkin Elmer, Walham, MA, USA) (λ_{ex} = 525, λ_{em} = 560). A standard curve was prepared by dissolving hydrolyzed TEP in water at different concentrations (33.5, 16.8, 8.4, 4.20, 2.10, 1.05, and 0.52 μM). The results have been expressed as nmol MDA/g tissue.

The protein carbonylation was determined using the method adapted from Terevinto and colleagues [19]. Liver and brain samples were homogenized and incubated with 0.02 M DNPH in 2 M HCl. Proteins were then precipitated by adding 20% TCA and recovered by centrifugation (625× g for 10 min). Pellets were washed three times with ethanol:ethyl acetate (1:1, *v/v*), melted in 6 M guanidine HCl in 0.02 M KH_2PO_4 (pH 6.5), and centrifuged. The absorbance of the supernatant was measured at 390 nm. The results have been expressed as nmol/g tissue.

The GSH content in liver and brain samples was evaluated according to Browne and Armstrong [20], with slight modifications. Proteins were then precipitated by adding 10% TCA (*w/v*) at 4 °C for 30 min. An aliquot of 150 μL of the sample was incubated with an equal volume of *o*–phthaldehyde (1 mg/mL) in 10% methanol (*v/v*) by 15 min at 37 °C. After centrifugation (625× g for 3 min), fluorescence was measured (Perkin Elmer LS-45, Perkin Elmer, Walham, MA, USA) (λ_{ex} = 350, λ_{em} = 420). A calibration curve has been performed by dissolving GSH in water at different concentrations (50, 25, 12.5, 6.25, 3.13, 1.56, 0.78 μM), and GSH concentrations have been calculated as μmol GSH/g tissue.

2.8. Statistical Analysis

The statistical analyses have been performed by Statistical Package for Social Science (SPSS) 17 for Windows (SPSS, Inc., Chicago, IL, USA). The results are shown as the mean value ± standard deviation (s.d.) and analyzed through a one-way ANOVA and Tukey's test for post-hoc with significance at $p \leq 0.05$.

3. Results and Discussion

3.1. Quantification of Total Polyphenols

The TPC of wild Italian blackthorn fruit extract was quantified by a spectrophotometric method, and the content was 5.50 ± 0.19 mg GAE/g d.w. To our knowledge, any other results have been found about total phenolic content of blackthorn fruit on dry weight, but some authors reported that TPC in blackthorn fruit on fresh weight ranged from 0.42–4.13 mg GAE/g [21–23].

The HPLC-DAD method validation was estimated by quantifying the limit of detection (LOD), limit of quantification (LOQ), and the recovery. All parameters indicate that the method exhibits a good sensitivity for identification as well as quantification of the polyphenols. All the parameters are listed in Table 1.

Table 1. Retention time (Rt), LOD (limit of detection), LOQ (limit of quantification), and recovery of phenolic compound quantification method by HPLC-DAD in the *Prunus spinosa* L. fruit (PSF) aqueous extract (*n* = 3).

Phenolic Compound	Rt (min)	LOD (ug/mL)	LOQ (ug/mL)	Recovery (%)
Gallic acid	2.860	0.012	0.033	98.2 ± 0.81
Rutin	5.909	0.009	0.030	89.1 ± 0.89
4-hydroxibenzoic acid	7.112	0.005	0.017	101.2 ± 1.01
Caffeic acid	8.361	0.008	0.027	97.5 ± 0.99
Trans p-coumaric acid	11.741	0.004	0.013	96.0 ± 0.80
Trans -ferulic acid	12.981	0.003	0.010	97.8 ± 1.08
Trans-sinapic acid	13.062	0.011	0.037	98.2 ± 1.15
Myricetin	17.081	0.015	0.050	99.5 ± 0.88
Rosmarinic acid	17.463	0.009	0.023	102.1 ± 0.96
Quercetin	18.853	0.090	0.299	99.5 ± 0.95
Genistein	19.811	0.009	0.031	91.5 ± 0.77

The quality of phenolic profile of wild Italian blackthorn and the concentrations of single compounds are shown in Table 2. Rutin (183.94 mg/kg d.w.) was the principal phenolic component, followed by 4-hydroxybenzoic acid, gallic acid, trans-sinapic acid, quercetin, trans-ferulic acid, caffeic acid, rosmarinic acid, trans cumaric acid, genistin, and myricetin. Our findings are partially in accordance with those of some other authors that showed considerable quantities of phenolic acids (quercetin and caffeic acid) in blackthorn fruits from Southeast Serbia [3,24]; by contrast we did not find either neochlorogenic or kaempferol. HPLC-UV analysis of the methanolic extract of fresh blackthorn plums from Turkey recently allowed Baltas and colleagues to detect five phenolic acids, namely protocatechuic acid, p-OH benzoic acid, vanillic acid, syringic acid, and p-coumaric acid, as well as flavonoids, such as epicatechin and luteolin [25]. Another recent study about quantification of phenolic compounds by HPLC-UV in methanolic extract of frozen blackthorn fruits from Romania showed high chlorogenic and neochlorogenic acid levels, followed by glycosides of quercetin [23]. Considering that the solubility of polyphenols in solvent of different polarity is determined by their structure, different types of extraction solvent and procedures may influence the efficiency of phenolic compounds extraction and their resultant content [3].

Table 2. Concentrations of phenolic compounds in the PSF aqueous extract.

Phenolic Compound	Concentration (mg/kg d.w.)
Gallic acid	41.10 ± 3.68
Rutin	183.94 ± 0.45
4-hydroxybenzoic acid	73.93 ± 0.06
Caffeic acid	3.36 ± 0.36
Trans *p*-coumaric acid	2.99 ± 0.02
Trans-ferulic acid	4.93 ± 0.07
Trans-sinapic acid	37.69 ± 0.05
Myricetin	1.47 ± 0.03
Rosmarinic acid	3.23 ± 0.03
Quercetin	9.94 ± 0.01
Genistin	1.74 ± 0.00

3.2. Antimicrobial Activity

The antimicrobial activity on selected Gram-negative (Figure 2A) and Gram-positive (Figure 2B) enteric bacteria was measured by evaluating the growth inhibition by increasing concentrations of PSF extract. The antimicrobial activities have been compared with the standard antibiotics, used as positive controls.

The lowest concentration of tested PSF extract (0.25 mg/mL) inhibited more than 50% of the Gram-negative bacteria *Escherichia coli* (70.19% ± 1.21%), *Salmonella typhimurium* (79.98% ± 0.54%), and *Enterobacter aerogenes* (83.02% ± 0.54%) growth (Figure 2A). The same concentration (0.25 mg/mL) was able to inhibit more than 50% of the Gram-positive bacteria *Enterococcus faecalis* (82.86% ± 1.94%) and *Staphylococcus aureus* (79.92% ± 1.23%) growth (Figure 2B). The antimicrobial activity of phenolic compounds occurring in plant foods has been widely studied against a wide range of microorganisms. The damage to the bacterial membrane and suppression of some virulence factors, including enzymes and toxins, are suggested to be the mechanisms of their antimicrobial action [26]. Some flavonoids (rutin, myricetin, and quercetin) and phenolic acids (gallic, caffeic, and ferulic acids) of PSF extract may be responsible for its antibacterial action [27,28].

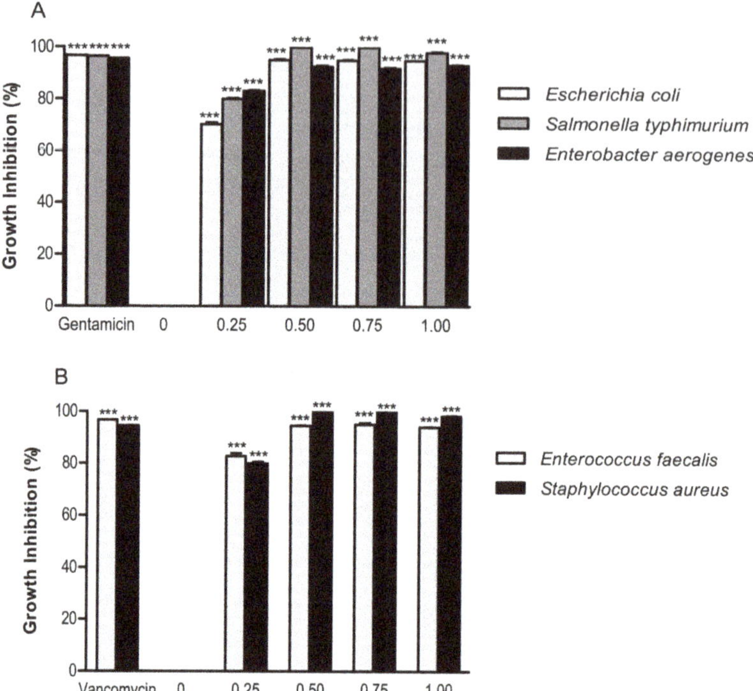

Figure 2. Growth inhibition effect of PSF extract (0, 0.25, 0.50, 0.75, and 1.00 mg/mL) against Gram-negative bacteria (**A**) (*Escherichia coli* ATCC 25922, *Salmonella enterica* ser. *typhimurium* ATCC 14028, and *Enterobacter aerogenes* ATCC 13048) and Gram-positive bacteria (**B**) (*Enterococcus faecalis* ATCC 29212 and *Staphylococcus aureus* ATCC 25923). Significantly different from negative control (PSF 0 mg/mL): *** $p \leq 0.001$. Results are reported as means ($n = 3$) values ± standard deviation.

3.3. In Vitro Antioxidant Activity

As shown in Figure 3A, pretreated erythrocytes with PSF aqueous extract (100 mg/mL) exhibited a significantly higher cellular antioxidant activity (CAA unit = 48.43 ± 1.68) compared with untreated cells (CAA = 0; $p \leq 0.001$), comparable to 100 μM Trolox (CAA unit = 16.52 ± 3.60; $p \leq 0.001$) and 500 μM Trolox (CAA unit = 36.67 ± 1.48; $p \leq 0.001$). Taking these results into consideration, the EC50 of PSF extract for antioxidant activity in red blood cells was 100 mg/mL.

The antihemolytic activity of PSF extract was screened in erythrocytes exposed to high doses of AAPH, causing a strong oxidative hemolysis. Figure 3B shows that PSF extract exerted a significant inhibition of AAPH-induced hemolysis compared with the control erythrocytes (AAPH-treated). PSF extract (100 mg/mL) pretreated cells demonstrated a marked antihemolytic effect (84% hemolysis inhibition) compared with AAPH-treated cells ($p \leq 0.001$), with a reduction of the hemolysis similar to that of the highest concentration of the reference standard (500 μM Trolox). The antihemolytic EC50 of PSF extract was 10 mg/mL (data not shown). We found that PSF exerted a potent ROS-scavenger activity. Indeed, when intact human erythrocytes were pre-incubated with a PSF aqueous extract, a strong protective effect against AAPH-generated ROS production and hemolysis was observed. These antioxidant and antihemolytic effects of PSF are probably due to the activity of gallic acid, rutin, and quercetin in red blood cell [29,30].

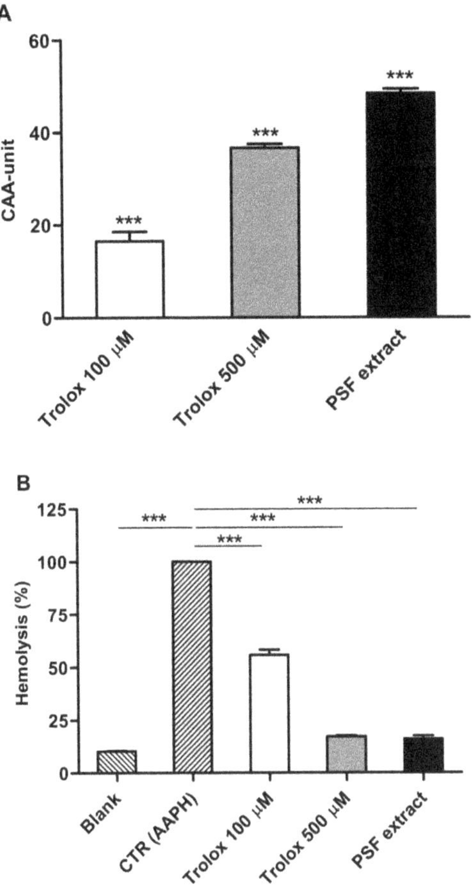

Figure 3. (**A**) Effects of PSF extract (100 mg/mL) on cellular antioxidant activity (CAA) in human erythrocytes. Significantly different from untreated cells (CAA unit = 0): *** $p \leq 0.001$. (**B**) Effects of PSF extract (100 mg/mL) on dihydrochloride (AAPH)-induced oxidative hemolysis in human erythrocytes. Significantly different from CTR (AAPH-treated cells): *** $p \leq 0.001$. Trolox was used as reference standard. Results are reported as means (n = 3) values ± standard deviation.

3.4. *In Vivo Experiment*

3.4.1. The Effect of Blackthorn on Body Weight and Liver Weight

In comparison with CTR group, rats of the HFD group exhibited a significant lower final body weight (396.8 ± 40.6 vs. 307.5 ± 23.3 g/rat, respectively) ($p \leq 0.001$). The administration of PSF did not induce significant changes in the final body weight, neither in PSF400 group (317.2 ± 27.6 g/rat), nor in PSF800 group (312.7 ± 42.5 g/rat), when compared with HFD group.

However, when compared with CTR rats, HFD rats exhibited a statistically significant increase in liver weight (8.9 ± 1.4 vs. 13.4 ± 1.2 g, respectively) ($p \leq 0.001$) and in relative liver weight (2.2 ± 0.2 vs. 4.1 ± 1.0 g liver/100 g b.w., respectively) ($p \leq 0.001$). No significant difference in liver weight was found between HFD-fed rats and PSF-treated rats of PSF400 group (14.8 ± 1.7 g) and PSF800 group (14.6 ± 1.8 g). The same trend was found in relative liver weight between HFD-fed rats and PSF-treated rats of PSF400 group (4.5 ± 0.3 g liver/100 g b.w.) and PSF800 group (4.8 ± 0.7 g liver/100 g b.w.). HFD treatment caused hepatic lipid accumulation and increased liver weight and all the biochemical parameters in serum [16,31]. However, the PSF extract did not improve the liver weight and serum and liver biochemical parameters linked to steatosis.

3.4.2. The Effect of PSF on Serum and Liver Biochemical Parameters

Serum AST, ALT, glucose, total cholesterol, triglycerides, and total hepatic lipid content were significantly higher in the HFD group compared with the normal diet group (CTR group), while serum insulin was significantly lower. After four weeks of treatment with 800 mg of PSF/kg b.w., rats of the PSF800 group showed a significant decrease of total hepatic lipids content compared with HFD group (162.15 ± 35.52 vs. 209.90 ± 11.91; $p \leq 0.05$) (Table 3). Some studies have demonstrated that polyphenols decrease the hepatic lipid accumulation caused by high-fat diet [32]. Moreover, it was also reported that the single isolated polyphenol can improve the high liver lipids content due to a high-fat diet administration, as in the case of rutin [33], gallic acid [34], and quercetin [35]. The crude extracts can be more advantageous than the isolated components, since a single bioactive molecule can change its properties with the presence of other compounds in the extracts [36].

Table 3. Nutritional effect of PSF on biochemical parameters in serum and liver of rats ($n = 7$).

	CTR	HFD	PSF400	PSF800
AST (UI/dl)	93.98 ± 7.04	194.33 ** ± 42.90	181.00 ** ± 22.23	168.50 * ± 44.81
ALT (UI/dl)	39.06 ± 10.09	143.17 ** ± 35.15	146.52 ** ± 51.96	133.58 ** ± 36.90
Insulin (µg/l)	1.44 ± 0.80	0.19 ** ± 0.15	0.22 ** ± 0.15	0.20 ** ± 0.03
Glucose (mg/dl)	145.20 ± 20.80	439.67 *** ± 70.21	432.57 *** ± 33.57	443.60 *** ± 43.32
Total cholesterol (mg/dl)	109.30 ± 21.40	236.20 *** ± 45.09	228.36 *** ± 29.13	219.18 *** ± 17.94
Triglycerides (mg/dl)	75.73 ± 7.38	179.80 ** ± 59.30	164.33 ** ± 30.52	170.25 ** ± 21.80
Total hepatic lipids (mg/g)	65.36 ± 9.14	209.90 *** ± 11.91	198.29 *** ± 34.52	162.15 *** § ± 35.52

Analyses were performed through one-way ANOVA and Tukey's test for post-hoc. *, $p \leq 0.05$ vs. CTR; **, $p \leq 0.01$ vs. CTR; ***, $p \leq 0.001$ vs. CTR; §, $p \leq 0.05$ vs. HFD. AST, aspartate aminotransferase; ALT, alanine aminotransferase.

3.4.3. The Effect of PSF on Liver and Brain Oxidative Stress of Rats

The high-fat diet was probably responsible for the decrease of GSH (Figure 4A) content and the increase of protein carbonylation (Figure 4C) and MDA (Figure 4E) levels in the liver samples of HFD rats, compared with the CTR group. Moreover, while hepatic GSH content was not affected by PSF treatment (Figure 4A), administration of PSF improved the oxidative stress status of rats according to the protein carbonylation, at the higher concentration of treatment (800 mg/kg b.w.) (Figure 4C), and to the MDA content, in a dose-dependent manner (Figure 4E). It has been demonstrated that plant polyphenols are related to the improvement of hepatic oxidative stress caused by a high-fat diet through the e activation of Nrf2 transcription factor, which increases expression of antioxidant enzymes [37]. Moreover, it was reported that even the single polyphenol, if isolated, can improve the high-fat-diet-induced hepatic oxidative stress, as in the case of rutin [33] and gallic acid [38].

In comparison with the CTR group, HFD treatment promoted an increase of the brain oxidative stress parameters in rats, as shown by protein carbonylation (Figure 4D) and MDA assay (Figure 4F). The addition of 400 mg/kg b.w. and 800 mg/kg b.w. of PSF to the diet reversed the effect caused by the high-fat diet and, in particular, the MDA assay showed a dose-dependent response pattern. Nevertheless, both the HFD and the PSF treatments did not induce significant changes in rat brain GSH levels (Figure 4B). The intake of a high-fat diet is linked to an increased risk of neurodegenerative disease related to diabetes [39]. Considering that, the polyphenols-rich fruits could protect neurons against the oxidative stress induced by intake of saturated fatty acids [40]. Recently, Nabavi and colleagues demonstrated that gallic acid exerts a neuroprotective effect against sodium fluoride-induced oxidative stress in rat brain [41]. Moreover, it has been shown that other polyphenols contained in PSF, such as rutin, ferulic acid, and trans-sinapic acid, should contribute to the prevention of brain oxidative stress in rats [42–44].

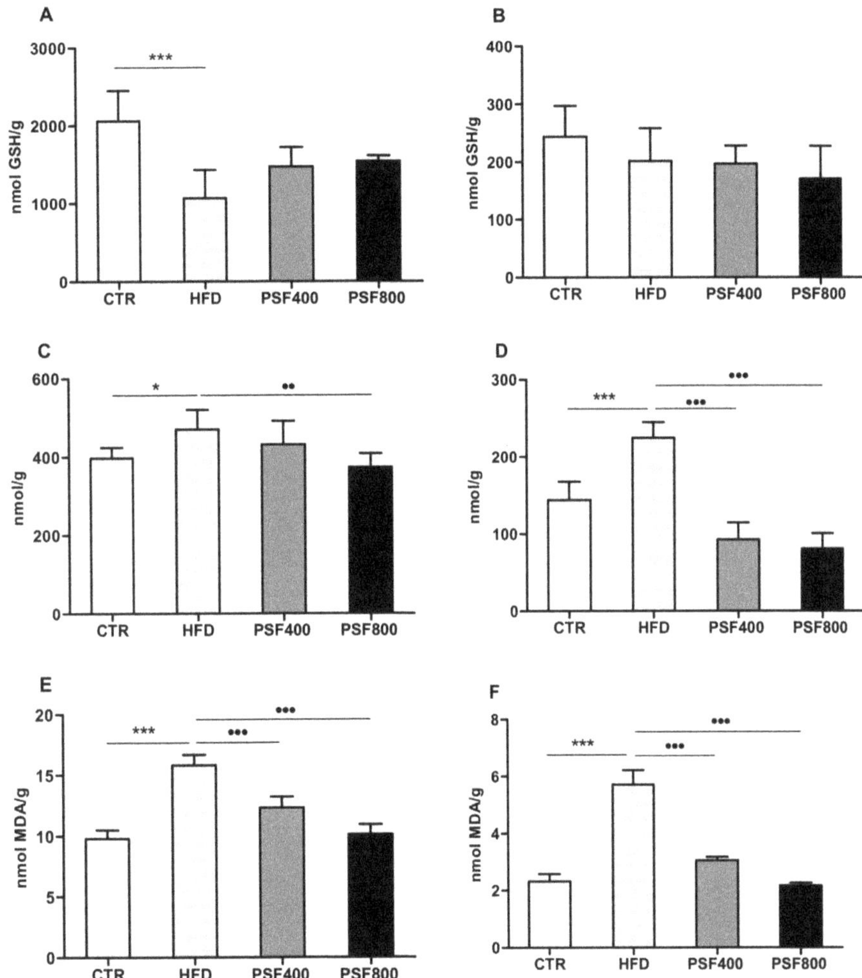

Figure 4. Effect of PSF treatment at two different concentrations (PSF400 and PSF800) on GSH content of liver (**A**) and brain (**B**), protein carbonylation of liver (**C**) and brain (**D**), and malondyaldeide of liver (**E**) and brain (**F**). Results are reported as means ($n = 7$) values ± standard deviation. Values within each row of different letters are significantly different ($p \leq 0.05$), $p \leq 0.05$ vs. CTR; **, $p \leq 0.001$ vs. CTR; ••, $p \leq 0.01$ vs. HFD; •••, $p \leq 0.001$ vs. HFD.

Our findings suggest an improved liver and brain antioxidant defense in rats treated with PSF.

4. Conclusions

All in all, our findings indicated that wild Italian blackthorn fruit is rich in polyphenol compounds, shows an in vitro antioxidant activity, and exhibits a selective growth inhibition of some potentially pathogenic bacteria strains. Moreover, this study is the first to evaluate an In Vivo antioxidant activity of PSF. In particular, our findings indicated that the oxidative stress arising in HFD group is decreased in liver and brain tissues by the intake of blackthorn fruit. The PSF supplementation demonstrated In Vivo antioxidant capacities, reducing liver and brain oxidative stress, probably due to the presence of polyphenols, such as rutin, 4-hydroxybenzoic acid, gallic acid, trans-sinapic acid, quercetin, trans-ferulic acid, caffeic acid, rosmarinic acid, trans coumaric acid, genistin, and myricetin, which were identified in the blackthorn fruit.

Thus, it is supposed that the regular consumption of wild Italian blackthorn fruit should increase the circulation of bioactive compounds, such as polyphenols, which could possibly improve the endogenous antioxidant system and protect tissues against oxidative stress damage induced by high-fat diet and hyperglycemia. Considering its beneficial properties, wild Italian blackthorn fruit can be potentially used to produce natural functional food, novel nutraceuticals, and it can also be employed in food processing.

Author Contributions: Investigation, L.P., R.R., F.V., A.V. and C.M.D.C.; Methodology, L.P., R.R., S.F., J.A., M.P. and C.M.D.C.; Supervision, D.C. and V.L.; Writing—original draft, L.P., R.R., F.V. and A.V. All authors have read and agreed to the published version of the manuscript.

Funding: This research received no external funding.

Acknowledgments: This study was supported by Institute of Biology and Agricultural Biotechnology of National Research Council (Italy), by Department of Agricultural, Environmental and Food Sciences, University of Molise (Italy) and by AgroBioTech Research Center (Slovakia), built in accordance with the project Building AgroBioTech Research Center ITMS 26220220180.

Conflicts of Interest: No conflict of interest was reported by the authors.

Abbreviations

ALT	alanine aminotransferase
ANOVA	analysis of variance
AST	aspartate aminotransferase aspartate aminotransferase
AATCC	american type culture collection
AAPH	2,2′-azobis (2-amidinopropane) dihydrochloride
CAA-RBC	cellular antioxidant activity in red blood cells
CFU	colony-forming unit
CTR	control
DCFH-DA	2,7-dichlorodihydrofluorescein diacetate
DNPH	dinitrophenylhydrazine
EC50	half maximal effective concentration
EDTA	ethylenediaminetetraacetic acid
GAE	gallic acid equivalent
GSH	reduced glutathione
HCl	hydrochloric acid
HFD	high fat diet
KH_2PO_4	potassium dihydrogen phosphate
LOD	limit of detection
LOQ	limit of quantification
MIC	minimum inhibitory concentrations
MDA	malondialdehyde
MHA	Mueller Hinton agar
MHB	Mueller Hinton broth
NA	nutrient agar
NB	nutrient broth
OD	optical density
PBS	phosphate buffer saline

References

1. Bhaskarachary, K.; Sudershan, R.V.; Subba Rao, M.G.; Apurva Kumar, R.J. Traditional foods, functional foods and nutraceuticals. *Proc. Indian Natn. Sci. Acad.* **2016**, *82*, 1565–1577.
2. Grochowicz, J.; Fabisiak, A.; Nowak, D. Market of functional food—legal regulations and development perspectives. *Zesz. Probl. Postepow Nauk Roln.* **2018**, *595*, 51–67. [CrossRef]

3. Veličković, J.M.; Kostić, D.A.; Stojanović, G.S.; Mitić, S.S.; Mitić, M.N.; Ranđelović, S.S.; Đorđević, A.S. Phenolic composition, antioxidant and antimicrobial activity of the extracts from *Prunus spinosa* L. fruit. *Hem. Ind.* **2014**, *68*, 297–303. [CrossRef]

4. Egea, I.; Sanchez-Bel, P.; Romorjaro, F.; Pretel, M.T. Six edible wild fruits as potential antioxidant additives or nutritional supplements. *Plant Foods Hum. Nutr.* **2010**, *65*, 121–129. [CrossRef] [PubMed]

5. Pinacho, R.; Yolanda Cavero, R.; Astiasarán, I.; Ansorena, D.; Calvo, M.I. Phenolic compounds of blackthorn (*Prunus spinosa* L.) and influence of In Vitro digestion on their antioxidant capacity. *J. Funct. Foods* **2015**, *19*, 49–62. [CrossRef]

6. Pandey, K.B.; Rizvi, S.I. Plant polyphenols as dietary antioxidants in human health and disease. *Oxid. Med. Cell Longev.* **2009**, *2*, 270–278. [CrossRef]

7. Vanzo, A.; Vrhovsek, U.; Tramer, F.; Mattivi, F.; Passamonti, S. Exceptionally fast uptake and metabolism of cyanidin 3-glucoside by rat kidneys and liver. *J. Nat. Prod.* **2011**, *74*, 1049–1054. [CrossRef]

8. Marchelak, A.; Owczarek, A.; Matczak, M.; Pawlak, A.; Kolodziejczyk-Czepas, J.; Nowak, P.; Olszewska, M.A. Bioactivity potential of *Prunus spinosa* L. flower extracts: Phytochemical profiling, cellular safety, pro-inflammatory enzymes inhibition and protective effects against oxidative stress In Vitro. *Front. Pharmacol.* **2017**, *8*, 680–695. [CrossRef]

9. Palaniappan, K.; Holley, R.A. Use of natural antimicrobials to increase antibiotic susceptibility of drug resistant bacteria. *Int. J. Food Microbiol.* **2010**, *140*, 164–168. [CrossRef]

10. Taleb, H.; Maddocks, S.E.; Morris, R.K.; Kanekanian, A.D. The antibacterial activity of date syrup polyphenols against *S. aureus* and *E. coli*. *Front. Microbiol.* **2016**, *7*, 1–9. [CrossRef]

11. Singleton, V.; Ortoger, R.M.; Lamuela-Ravendo, R.M. Analysis of total phenols and others oxidation substrates and antioxidants by means of Folin-Ciocalteau reagent. *Methods Enzymol.* **1999**, *299*, 152–173.

12. International Conference on Harmonization of Technical Requirements for Registration of Pharmaceuticals for Human Use. Validation of Analytical Procedures: Text and Methodology, Q2 (R1), 2005. Available online: https://www.ema.europa.eu/en/documents/scientific-guideline/ich-q-2-r1-validation-analytical-procedures-text-methodology-step-5_en.pdf (accessed on 2 December 2019).

13. Delgado-Adámez, J.; Fernández-León, M.F.; Velardo-Micharet, B.; González-Gómez, D. In Vitro assays of the antibacterial and antioxidant activity of aqueous leaf extracts from different *Prunus salicina* Lindl. cultivars. *Food Chem. Toxicol.* **2012**, *50*, 2481–2486. [CrossRef] [PubMed]

14. Frassinetti, S.; Gabriele, M.; Caltavuturo, L.; Longo, V.; Pucci, L. Antimutagenic and antioxidant activity of a selected lectin-free common bean (*Phaseolus vulgaris* L.) in two cell-based models. *Plant Foods Hum. Nutr.* **2015**, *70*, 35–41. [CrossRef] [PubMed]

15. Wolfe, K.L.; Liu, R.H. Cellular antioxidant activity (CAA) assay for assessing antioxidants, foods, and dietary supplements. *J. Agric. Food Chem.* **2007**, *55*, 8896–8907. [CrossRef] [PubMed]

16. Pozzo, L.; Vornoli, A.; Coppola, I.; Della Croce, C.M.; Giorgetti, L.; Gervasi, P.G.; Longo, V. Lipid, cholesterol and glucose metabolism gene expression in a high-fat-diet and low-dose-streptozotocin induced rat model. *Life Sci.* **2016**, *166*, 149–156. [CrossRef] [PubMed]

17. Folch, J.; Lees, M.; Sloane Stanley, J.H. A simple method for the isolation and purification of total lipids from animal tissues. *J. Biol. Chem.* **1957**, *226*, 497–509.

18. Seljeskog, E.; Hervig, T.; Mansoor, M.A. A novel HPLC method for the measurement of thiobarbituric acid reactive substances (TBARS). A comparison with a commercially available kit. *Clin. Biochem.* **2006**, *39*, 947–954. [CrossRef]

19. Terevinto, A.; Ramos, A.; Castroman, G.; Cabrera, M.C.; Saadoun, A. Oxidative status, In Vitro iron-induced lipid oxidation and superoxide dismutase, catalase and glutathione peroxidase activities in rhea meat. *Meat Sci.* **2010**, *84*, 706–710. [CrossRef]

20. Browne, R.W.; Armstrong, D. Reduced glutathione and glutathione disulfide. *Methods Mol. Biol.* **1998**, *108*, 347–352.

21. Kim, D.O.; Chun, O.K.; Kim, Y.J.; Moon, H.Y.; Lee, C.Y. Quantification of polyphenolics and their antioxidant capacity in fresh plums. *J. Agric. Food Chem.* **2003**, *51*, 6509–6515. [CrossRef]

22. Erturk, Y.; Ercisli, Y.; Tosum, M. Physico-chemical characteristics of wild plum fruits (*Prunus spinosa* L.). *Int. J. Plant Prod.* **2009**, *3*, 89–92.

23. Varga, E.; Domokos, E.; Fogarasi, E.; Ştefănesu, R.; Fü-löp, I.; Croitoru, M.D.; Laczkó-Zöld, E. Polyphenolic compounds analysis and antioxidant activity in fruits of Prunus spinosa L. *Acta Pharm. Hung.* **2017**, *87*, 1–7.

24. Radovanović, B.C.; Anđelković, S.M.; Radovanović, A.B.; Anđelković, M.Z. Antioxidant and antimicrobial activity of polyphenol extracts from wild berry fruits grown in Southeast Serbia. *Trop. J. Pharm. Res.* **2013**, *12*, 813–819. [CrossRef]

25. Baltas, N.; Parkyldiz, S.; Can, Z.; Dincer, B.; Kolayli, S. Biochemical properties of partially purified polyphenol oxidase and phenolic compounds of *Prunus spinosa* L. subsp. *Dasyphylla* as measured by HPLC-UV. *Int. J. Food Prop.* **2017**, *20*, 1377–1391. [CrossRef]

26. Barbieri, R.; Coppo, E.; Marchese, A.; Daglia, M.; Sobarzo-Sánchez, E.; Nabavi, S.F.; Nabavi, S.M. Phytochemicals for human disease: An update on plant-derived compounds antibacterial activity. *Microbiol. Res.* **2017**, *196*, 44–68. [CrossRef] [PubMed]

27. Hendra, R.; Ahmad, S.; Sukari, A.; Shukor, M.Y.; Oskoueian, E. Flavonoid analyses and antimicrobial activity of various parts of *Phaleria macrocarpa* (Scheff.) Boerl fruit. *Int. J. Mol. Sci.* **2011**, *12*, 3422–3431. [CrossRef]

28. Saavedra, M.J.; Borges, A.; Dias, C.; Aires, A.; Bennett, R.N.; Rosa, E.S.; Simões, M. Antimicrobial activity of phenolics and glucosinolate hydrolysis products and their synergy with streptomycin against pathogenic bacteria. *Med. Chem.* **2010**, *6*, 174–183. [CrossRef]

29. Asgary, S.; Naderi, G.H.; Askari, N. Protective effect of flavonoids against red blood cell hemolysis by free radicals. *Exp. Clin. Cardiol.* **2005**, *10*, 88–90.

30. Hatia, S.; Septembre-Malaterre, A.; Le Sage, F.; Badiou-Bénéteau, A.; Baret, P.; Payet, B.; Lefebvre d'hellencourt, C.; Gonthier, M.P. Evaluation of antioxidant properties of major dietary polyphenols and their protective effect on 3T3-L1 preadipocytes and red blood cells exposed to oxidative stress. *Free Radic. Res.* **2014**, *48*, 387–401. [CrossRef]

31. Vornoli, A.; Pozzo, L.; Della Croce, C.M.; Gervasi, P.G.; Longo, V. Drug metabolism enzymes in a steatotic model of rat treated with a high fat diet and a low dose of streptozotocin. *Food Chem. Toxicol.* **2014**, *70*, 54–60. [CrossRef]

32. Cheng, D.M.; Pogrebnyak, N.; Kuhn, P.; Poulev, A.; Watweman, C.; Rojas-Silva, P.; Johnson, W.D.; Raskin, I. Polyphenol-rich rutgers scarlet lettuce improves glucose metabolism and liver lipid accumulation in diet-induced obese C57BL/6 mice. *Nutrition* **2014**, *30*, 52–58. [CrossRef] [PubMed]

33. Al-Rejaie, S.S.; Aleisa, A.M.; Sayed-Ahmed, M.M.; AL-Shabanah, O.A.; Abuohashish, H.M.; Ahmed, M.M.; Al-Hosaini, K.A.; Hafez, M.M. Protective effect of rutin on the antioxidant genes expression in hypercholestrolemic male Wistar rat. *BMC Complement. Altern. Med.* **2013**, *13*, 136–145. [CrossRef] [PubMed]

34. Gandhi, G.R.; Jothi, G.; Antony, P.J.; Balakrishna, K.; Paulraj, M.G.; Ignacimuthu, S.; Stalin, A.; Al-Dhabi, N.A. Gallic acid attenuates high-fat diet fed-streptozotocin-induced insulin resistance via partial agonism of PPARgamma in experimental type 2 diabetic rats and enhances glucose uptake through translocation and activation of GLUT4 in PI3K/p-Akt signaling pathway. *Eur. J. Pharmacol.* **2014**, *745*, 201–216. [PubMed]

35. Ting, Y.; Chang, W.T.; Shiau, D.K.; Chou, P.H.; Wu, M.F.; Hsu, C.L. Antiobesity efficacy of quercetin-rich supplement on diet-induced obese rats: Effects on body composition, serum lipid profile, and gene expression. *J. Agric. Food Chem.* **2018**, *66*, 70–80. [CrossRef] [PubMed]

36. Lee, O.H.; Lee, B.Y. Antioxidant and antimicrobial activities of individual and combined phenolics in *Olea europaea* leaf extract. *Bioresour. Technol.* **2010**, *101*, 3751–3754. [CrossRef] [PubMed]

37. Tuzcu, Z.; Orhan, C.; Sahin, N.; Juturu, V.; Sahin, K. Cinnamon polyphenol extract inhibits hyperlipidemia and inflammation by modulation of transcription factors in high-fat diet-fed rats. *Oxid. Med. Cell Longev.* **2017**, *31*, 150–165. [CrossRef] [PubMed]

38. Karimi-Khouzani, O.; Heidarian, E.; Amini, S.A. Anti-inflammatory and ameliorative effects of gallic acid on fluoxetine-induced oxidative stress and liver damage in rats. *Pharmacol. Rep.* **2017**, *69*, 830–835. [CrossRef]

39. Winocur, G.; Greenwood, C.E. Studies of the effects of high fat diets on cognitive function in a rat model. *Neurobiol. Aging* **2005**, *26*, 46–49. [CrossRef]

40. Charradi, K.; Mahmoudi, M.; Bedhiafi, T.; Kadri, S.; Elkahoui, S.; Limam, F.; Aouani, E. Dietary supplementation of grape seed and skin flour mitigates brain oxidative damage induced by a high-fat diet in rat: Gender dependency. *Biomed. Pharmacother.* **2017**, *87*, 519–526. [CrossRef]

41. Nabavi, S.F.; Habtemariam, S.; Jafari, M.; Sureda, A.; Nabavi, S.M. Protective role of gallic acid on sodium fluoride induced oxidative stress in rat brain. *Bull. Environ. Contam. Toxicol.* **2012**, *89*, 73–77. [CrossRef]

42. Nkpaa, K.W.; Onyeso, G.I. Rutin attenuates neurobehavioral deficits, oxidative stress, neuro-inflammation and apoptosis in fluoride treated rats. *Neurosci. Lett.* **2018**, *24*, 92–99. [CrossRef] [PubMed]

43. Bais, S.; Kumari, R.; Prachar, Y. Ameliorative effect of trans-sinapic acid and its protective role in cerebral hypoxia in aluminium chloride induced dementia of Alzheimer's type. *CNS Neurol. Disord. Drug Targets* **2018**, *17*, 144–154. [CrossRef] [PubMed]
44. Asano, T.; Matsuzaki, H.; Iwata, N.; Xuan, M.; Kamiuchi, S.; Hibino, Y.; Sakamoto, T.; Okazaki, M. Protective effects of ferulic acid against chronic cerebral hypoperfusion-induced swallowing dysfunction in rats. *Int. J. Mol. Sci.* **2017**, *18*, 550. [CrossRef] [PubMed]

Article

Nutrient Composition of Popularly Consumed African and Caribbean Foods in The UK

Tanefa A. Apekey [1],*, June Copeman [1], Nichola H. Kime [2], Osama A. Tashani [1], Monia Kittaneh [1], Donna Walsh [1] and Maria J. Maynard [1]

[1] School of Clinical & Applied Sciences, Leeds Beckett University, Leeds LS1 3HE, UK;
    J.Copeman@leedsbeckett.ac.uk (J.C.); O.Tashani@leedsbeckett.ac.uk (O.A.T.); mkittaneh5@gmail.com (M.K.);
    d.walsh068@yahoo.com (D.W.); M.Maynard@leedsbeckett.ac.uk (M.J.M.)
[2] School of Sport, Leeds Beckett University, Leeds LS6 3QQ, UK; N.Kime@leedsbeckett.ac.uk
* Correspondence: T.A.Apekey@leedsbeckett.ac.uk; Tel.: +4-(0)113-812-4991

Received: 27 September 2019; Accepted: 11 October 2019; Published: 15 October 2019

Abstract: (1) Background: Traditional foods are important in the diets of Black Africans and Caribbeans and, more widely, influence UK food culture. However, little is known about the nutritional status of these ethnic groups and the nutrient composition of their traditional foods. The aim was to identify and analyse African and Caribbean dishes, snacks and beverages popularly consumed in the UK for energy, macronutrients and micronutrients. (2) Methods: Various approaches including focus group discussions and 24-h dietary recalls were used to identify traditional dishes, snacks, and beverages. Defined criteria were used to prioritise and prepare 33 composite samples for nutrient analysis in a UK accredited laboratory. Quality assurance procedures and data verification were undertaken to ensure inclusion in the UK nutrient database. (3) Results: Energy content ranged from 60 kcal in Malta drink to 619 kcal in the *shito* sauce. Sucrose levels did not exceed the UK recommendation for adults and children. Most of the dishes contained negligible levels of *trans* fatty acid. The most abundant minerals were Na, K, Ca, Cu, Mn and Se whereas Mg, P, Fe and Zn were present in small amounts. (4) Conclusion: There was wide variation in the energy, macro- and micronutrients composition of the foods analysed.

Keywords: nutrients; food composition; African; Caribbean; macronutrients; energy; vitamins and minerals

---

1. Introduction

In the United Kingdom (UK), as in other high-income countries, nutrition related ill-health is more common in some minority ethnic groups. For example, obesity, type 2 diabetes, hypertension and cardiovascular conditions are more common among ethnic groups of Black African origin compared to the majority population [1–4].

The development and implementation of effective public health programmes and nutrient recommendations requires reliable data on the nutritional status of the target population. In the UK, as elsewhere, national diet and nutrition surveys are regularly carried out in order to assess the dietary habits and nutritional status of the population. Information from these surveys inform government policies, public health education and interventions to promote nutrition related health and prevent non-communicable diseases [5]. According to the 2011 UK census, about 20% of the population self-identified as non-white British. Those from South Asian groups make up the largest minority ethnic group (7.5%). People of Black ethnicity were the second largest minority group (3.3%), with Black Africans being the fastest growing minority population [6,7]. The UK National Diet and Nutrition Survey (NDNS) which began in 1992 is designed to assess the dietary habits and nutritional status of adults and children [5]. The survey is the only source of high-quality data on dietary intakes and nutritional status in a representative sample of the population [8]. However, minority ethnic groups are not represented in the NDNS and other annual health surveys such as Health Survey for England [5,9,10]. To date, only two national health surveys have been conducted with boosted ethnic minority samples; this was in 1999 and 2004 [9,10]. The data collected involved questionnaire-based interviews, physical measurements, blood sample analysis, health and psychosocial wellbeing, cardiovascular disease (CVD) risk, tobacco use, alcohol consumption, obesity, blood pressure and physical activity and eating habits among the African-Caribbean, South Asian, Chinese and Irish groups throughout England. The data on eating habits was based on a food frequency questionnaire which did not include the traditional foods of these minority ethnic groups. The absence of these traditional foods is mainly due to the lack of reliable and comprehensive data on their nutrient composition. Ethnic foods are becoming increasingly popular and also contribute to the UK food culture They contribute around 19% of foods consumed (at least 4% of which are African and Caribbean foods) [11,12]. A reliable nutrient composition database of these traditional foods is therefore needed for comprehensive assessment of the nutritional status and dietary habits of these population groups. Accurate nutrient data are also essential in monitoring health and nutritional status as well as the development of tailored initiatives to tackle the widening inequalities in health and to improve nutrition related health [13].

The aim of the current study was to identify and analyse African and Caribbean dishes, snacks and beverages popularly consumed in the UK for energy, macronutrients and micronutrients. These new nutrient composition data will have various uses including nutritional surveys and health surveillance in Black ethnic groups and the majority population of the UK. This study is part of the programme of research of the Migrant Health Research group, School of Clinical and Applied Sciences, Leeds Beckett University. One of the aims of the group is to develop reliable and comprehensive nutrient composition data for popular multi-ethnic foods in the UK.

2. Materials and Methods

The full details of the methodology have been described elsewhere [14]. The procedures followed in developing these data are in line with the FAO (Food and Agriculture Organisation of the United Nations) and INFOODS (International Network of Food Data Systems) guidelines on production, management and data quality of food composition data [15,16].

Briefly, all volunteers were provided with an information sheet on the study and written consent was obtained; in accordance with the 1975 Declaration of Helsinki. The study was approved by the Faculty of Health and Social Science Research Ethics Committee, Leeds Beckett University (reference number 22,946).

2.1. Identification, Selection and Sampling of Popularly Consumed African and Caribbean Foods

Different sources including Mintel reports on ethnic foods and restaurants in UK [17,18], consumption data from food surveys and research papers [19–31] as well as data from major food retailers including ethnic food retailers, manufacturers, restaurants and takeaways were used to identify popularly consumed North African, West African and Caribbean dishes, snacks and beverages in the UK. Additional new data were collected using 24-h dietary recall, 10 focus group discussions and 5 individual interviews with African (North and West) and Caribbean adult over 18 years, living in Leeds, UK. See Figure S1 for stages involved in the selection of dishes, snacks and beverages for analyses.

A total of 33 (14 West African, 14 Caribbean and 5 North African) dishes, snacks and beverages were prioritised for nutrient analyses. Prioritisation was based on food consumption patterns, common nutrition-related diseases, consumer demand and preference, relevance to health inequalities and data from the focus groups and individual interviews. Traditional desserts are not commonly consumed [19–31] and therefore were not included. Table 1 shows the description of the 33 prioritised foods (dishes ($n = 26$), snacks ($n = 3$, plantain chips, meat patties and fried dumplings) and dessert ($n = 1$, *kunafa*) and beverages ($n = 3$, 'Malta' or other malt drink, rum and Guinness (Irish stout beer) punch).

65

Table 1. Description and proportion of ingredients for the prioritised dishes, snacks and beverages.

Ethnic Group	Food	Food Description
1. Caribbean	Rice and peas	Rice (19%) boiled in water (26%) and combined with black eyed, split or pigeon peas or kidney beans (26%); onions (0.5%), vegetable oil (2%), salt (0.5%) and coconut cream (26%) may be added.
2. Caribbean	Ackee and saltfish	Tinned ackee 38% (a tropical fruit, yellow in colour), saltfish (40%), onion (0.5%), garlic (0.5%), red/yellow pepper (0.5%), chopped tomato (19%), curry power/jerk seasoning (0.5%) and spring onion (0.5%).
3. Caribbean	West Indian soup	Made with meat (14%), dumplings (14%), large pieces of vegetables such as yam (7%), sweet potato (7%), pumpkin (7%), carrots (7%), noodle (7%) and chocho (a green tropical fruit, 11%) in thin stock, curry (5%) and water (21%).
4. Caribbean	Goat curry	Goat meat (65%) usually seasoned overnight with curry powder (2%), onions (0.5%), ginger (1%), cloves (0.5%) and scotch bonnet chillies (1 to 2) and then fried in oil (10%), water (20%) is added and left to cook until tender. Coconut cream (18%) and tomato puree (2%) may be added, with less water.
5. Caribbean	Jerk chicken	Chicken wings (80%), onions (2.5%), pepper (2.5%) and jerk sauce (marinade made with hot spices, 15%).
6. Caribbean	Caribbean fish curry	Headless red/white fish/ haddock (58%), coconut milk (20%), garlic (0.5), thyme (0.5%), carrots (2%), curry powder (2%), tomato (11%), spring onion (2%), onions (2%) and knob of butter (2%).
7. Caribbean	Caribbean vegetable curry	Red/white onion (2%), broccoli (8%), courgetti (8%), butternut squash (8%), cauliflower (8%), carrots (8%), green beans (8%), aubergines (8%), tomato (8%), thyme (0.5%), coconut milk (30%), curry powder (2.5%) and knob of butter (1%).
8. Caribbean	Callaloo and saltfish	Tinned callaloo (a green tropical fruit, 45%), saltfish (28%), water (16%), onion (2%), garlic (2%), carrots (3%), red/yellow pepper (2%) and spring onion (2%).
9. Caribbean	Cornmeal porridge	Hot milk (71%) and cornmeal flour (24%) (about 1.4% sweetened condensed milk may be added) flavoured with fresh nutmeg (0.3%), salt (0.3%), sugar (1%) and vanilla (2%). Cinnamon sticks or powder may also be added for flavour.
10. Caribbean	Guinness punch	Guinness (Irish stout beer, 36%), sweetened condensed milk (21%), vanilla extract (0.4%), cinnamon (0.3%), whole milk (42%) and nutmeg (0.3%).
11. Caribbean	Rum punch	White rum (18%), dark rum (10%), syrup (1.5), lemon/lime (4.5%), water (33%) and pineapple juice (33%).
12. Caribbean	Fried dumplings Also called 'Johnny cake'	Deep-fried (oil 20%) dough made with white flour (40%), nutmeg or vanilla extract (0.5), sugar (3%), butter (3%), salt (0.5%), water (32%) and baking powder (1%). Cornmeal may be added.
13. Caribbean	Saltfish fritters	Deep fried batter with saltfish/salted cod (30%), which is purchased dried and soaked overnight to remove salt or boiled to rehydrate, self-raising flour (23%), Scotch bonnet chilli (1%) and cooking oil (46%).
14. Caribbean	Meat patties	Semi-circular or oval shaped pastry [made with all-purpose flour (35%), water (9%), butter/vegetable shortening (5%), salt (0.2%)] filled with seasoned minced beef [made with ground beef (45%), chilli pepper (2%), onion (2%), thyme (0.8%) and garlic (1%)]. Vegetables may be added.
15. West African	Kenkey*◊	Fermented corn dough made into a ball, wrapped in corn husk and cooked over heat.
16. West African	Shito sauce ◊	Chilli/spicy Ghanaian sauce made with vegetable oil, onion, ginger, tomatoes, dried chilli, smoked fish, smoked shrimps, stock cube and spices.
17. West African	Cassava and plantain fufu*◊	Cassava, plantain and potato flour which may contain E102, E110, E450, E471 and/or E321
18. West African	Malt/Malta drink◊	Water, barley malt, glucose syrup/sugar, carbon dioxide, colour (E150c), acid (citric acid), liquorice, nicotinamide, pantothenol, thiamin hydrochloride, sodium, riboflavin, phosphate and pyridoxin chloride
19. West African	Plantain chips (chilli and plain) ◊	Ripe plantain, vegetable oil, sea salt, powdered chilli, spices, citric acid as a flavour enhancer.
20. West African	Eba (also known as Gari)	Ground cassava (80%) and water (20%).

Table 1. *Cont.*

Ethnic Group	Food	Food Description
21. West African	Rice and peas/beans	Black eye/brown bean (26%), long grain/basmati rice (20%), salt (1%) and water (53%).
22. West African	Jollof rice	Long grain/basmati rice (40%), tomatoes (27%), vegetable oil (5%), salt (0.5%), beef and/or chicken stock (10%), chicken/beef (10%), curry (1%), thyme (1%), onion (1%), ginger (0.5%), carrot (2%), Maggi chicken cube (0.5%), garlic (0.5%) and hot red pepper, or scotch bonnet/chilli pepper (1%). Other vegetables, chicken and/or beef may be added.
23. West African	Egushi stew	Egushi (ground melon seeds, 20%), beef stock (5%), stock fish (8%), dried fish (8%), beef (8%), salt (0.2), onions (1%), ugu leaf (fluted pumpkin leaves, 9.6%), Maggi chicken cube (0.2), palm oil (5%), garlic and tomato (35%).
24. West African	Groundnut soup	Peanut butter (15%), tomato (30%), tomato puree (0.8%), scotch bonnet (0.1%), Maggi chicken cube (0.3%), beef (15%), goat (15%), fish (15%), ginger (0.2%), vegetables (7.4%, optional - okro/okra, garden eggs), onions (0.8%) and salt (0.4%).
25. West African	Meat stew	Tomato (43%), tomato puree (9%), beef (37%), vegetable oil (6%), Maggi cube (0.5%), salt (0.5%), scotch bonnet/chilli pepper (0.2%), onions (1%), curry (2%), thyme (0.4%) and ginger (0.4%).
26. West African	One pot pepper/light soup	Soup prepared with vegetables (tomato (20%), tomato puree (4%), scotch bonnet (0.2%), ginger (0.3%), garlic (0.3%)], meat (15%), goat (15%), chicken (15%) and fish (15%), salt (0.2%), cow foot (5%), Maggi stock cube (0.2%), thyme (0.2%) [optional—okra/okra (4.8%) and/or garden egg (4.8%) also known as eggplant)
27. West African	Okro soup/stew	Okro (32%), tomato (21%), scotch bonnet/chilli pepper (0.2%), ginger (0.3%), garlic (0.2%), palm/vegetable oil (6%), beef (17.5%), fish (17.5%), spinach (5%), Maggi stock cube (0.3%).
28. West African	Ewedu soup	Ewedu leaves (33%, jute leaves), cray fish (5.8%), Maggi stock cube (0.5%), salt (0.5%), water (60%) and powdered potash (0.2%).
29. North African	Couscous with chicken	Couscous (25%), chicken (35%), water (20%), onion (0.5%), oil (3%), tomato (5%), mixed spices (5%), carrots (3%), chilli pepper (0.3%), coriander (0.2%) and chickpea (3%).
30. North African	Couscous with vegetables	Onion (1%), chickpea (3%), tomato (2%), tomato paste (0.8%), mixed spices (1.3%), water (26%), chilli pepper (0.1%), salt (0.2%), parsley/parsley (0.3%), oil (3%), potato (5.3%), carrots (7.8%), cabbage (7.8%), turnip (7.8%), butternut (7.8%), pumpkin (7.8%), couscous and (18%).
31. North African	Couscous with lamb	Onion (0.5%), lamb (36%), chilli pepper (0.3%), butternut (5%), parsley (0.2%), ghee (4%), tomato paste (3%), potato (5%), carrots (5%), squash (5%), couscous (31%), water and mixed spices (5%).
32. North African	Traditional Libyan soup (*Sharba Libiya*)	Lamb (24%), onion (1%), tomato (2%), vegetable oil/ghee (14%), herbs (6%), tomato paste (1%), soup pasta (22%), salt, chickpea (23%), cinnamon sticks (0.5%), cardamom pods (0.5%), bay leaf (0.5%) and mixed spices (5.6%).
33. North African	Kunafa (sugar-soaked pastry)	Kunafa pastry(22%), walnuts (20%), butter (9%), sugar (15%), raisins (5%), water (10%), lemon (0.2%), vanilla (0.3%), cinnamon (0.5%) and extra thick cream (18%)

Food is used in the table to represent dishes, beverages or snacks. *Modified foods i.e., ingredients and or recipes and cooking methods modified to align with UK tastes. ◊ Dishes, snacks and beverages did not require cooking.

2.2. Sampling, Preparation and Analyses of Prioritised Dishes, Snacks and Beverages

Traditional foods and ingredients were purchased from the four UK supermarkets with the largest market share (Tesco, Sainsbury's, Asda and Morrison's) as well as ethnic food shops and stalls, by stratified sampling approach. Stratification was based on type of retail outlet or sale point, sources, location and manufacturer brands. Ingredients were randomly purchased within each stratum in order to account for variations such as manufacturer's brands, processing conditions and retail outlet.

Female volunteers, 6 West African, 6 North African and 9 Caribbean were recruited to cook the prioritised dishes, beverages and snacks in the Nutrition kitchen at the University. They were recruited from places of worship and recreation, and other local hubs through word of mouth and poster advertisements. They all received an information sheet on the study and written consent was obtained.

The volunteer cooks regularly prepare and consume their assigned traditional dishes, beverages and snacks, as such were familiar with the ingredients, recipes and cooking procedures. Prior to the cooking sessions, the volunteers provided the list of ingredients and recipes, including quantities. Recipe harmonisation was by identification of common recipes, types and quantity of ingredients and methods of food preparation from the sources previously mentioned. These recipes and ingredients matched those provided by the volunteers. Preparation of dishes, snacks and beverages was therefore based on the harmonised recipes.

Composite samples were prepared according to procedures described by Apekey et al. [14]. Equal weights (500 g of edible portions) of similar foods, beverages or snacks were combined by mixing in a food blender to form a composite sample weighing $\approx$ 4000 g. Composite samples were prepared from 1 to 8 primary samples in order to reflect the variability in the composition due to recipe variations. Rigorous quality assurance procedures and verification of data were undertaken to ensure inclusion in the UK nutrient database, McCance and Widdowson's The Composition of Foods. A total of 33 samples were sent to a UK accredited laboratory in Leeds for nutrient analyses. See Figure S2 and Figure S3 for composite sample preparation process.

The methods used are accredited through the United Kingdom Accreditation Service (UKAS) to the ISO 17025 (International Organisation for Standardisation) standard and as such are fully validated. In order to meet the repeatability criteria documented in the methods used, the analytical tests were repeated. The analytical methods used are described in Table 2.

Table 2. Analytical methods used for the nutrient analysis.

Nutrient	Reference Method
Ash	BS4401-1 1998 ISO 936:1998
Moisture	BS4401-3:1997 ISO 1442:1997
Nitrogen (Total nitrogen)	Elementar Rapid N Cube Condensed Manual
Fatty acids by FAME Profile (MUFA, PUFA, SFA, Trans)	Kirk, R S, Sawyer, R, Pearson's Composition and Analysis of Foods, 9th edn, Longman, 1991, p24
Sugars	The sample is dissolved in water, with heating, clearing and dilution if necessary, and analysed by high performance liquid chromatography using refractive index detection
Chloride	The test sample is extracted in hot water, filtered and analysed by ion chromatography
K, Ca, Mg, P, Fe, Cu, Zn, Cl, Mn, Se	The samples are digested using a digiprep digestion block and analysed by ICP-MS
Sodium	The sample is ashed, the ash dissolved in water and the sodium content determined by Flame Photometry
Dietary fibre (AOAC)	AOAC method 985 29
Dietary fibre (NSP)	The sample undergoes enzymatic hydrolysis of starch, precipitation of NSP in ethanol, acid hydrolysis of the NSP and measurement of the released constituent sugars (Englyst)
Energy	Calculated from the protein, fat, carbohydrate (including sugars), and AOAC fibre using the values in Annex XIV of Regulation (EU) No 1169/2011. No allowance was made for the presence of any polyols, salatrims, alcohol, organic acid or erythritol
Protein	Total nitrogen multiplied by 6.25
Total fat	BS:4401:Part 4 1970

Table 2. *Cont.*

Nutrient	Reference Method
Carbohydrate	Available carbohydrate is calculated by difference (100 minus the sum of protein, total fat, ash, moisture, alcohol and AOAC fibre) in 100 g of food
Vitamin A	A7272: Vitamin A (Retinol). EN 12823-1 2014, LC-DAD
ß carotene	A7270: Beta-carotene, juice and vegetables. Pro-vitamin A; EN 12823-2:2000, LC-DAD
Vitamin B1 thiamine base	A7273: Vitamin B1 -Thiamine base. EN 14122:2003, mod., LC-FLD, Food and feed
Vitamin B12	DJCDE: Vitamin B12 HPLC (Immuno) Food and Feed. J. AOAC 2008, vol 91 No 4, LC-UV/DAD
Vitamin B2 (riboflavin)	A7274: Vitamin B2—riboflavin; EN 14152:2003, mod. LC-FLD
Vitamin B5 (Panthotenic acid)	DJ5BG: Vitamin B5 LC/MS/MS; AOAC 2012.16, LC/MS/MS with isotope dilution.
Biotin (vitamin H)	A7284: Vitamin B8—biotin, microbiological. Analogous to FDA method, LST AB 266.1,1995, Nephelometry
Vitamin B6 (pyridoxine)	A7251: Vitamin B6. EN 14164, LC-FLD
Calcium Pantothenate	See Vitamin B5 (Panthotenic acid)
Vitamin C	A7291: Vitamin C. Food Chemistry, 94 626-631, LC-DAD
Vitamin D2	A7294: Vitamin D2 (µg/100g). EN 12821:2009, LC-DAD
Folic acid total (vitamin B9/M)	A7286: Vitamin B9—Total folate, microbiological; NMKL 111:1985, Nephelometry
α, β, γ and δ Tocopherol	See sum of tocopherols
Sum tocopherols	A7297: Vitamin E (tocopherol profile). EN 12822:2014, LC-FLD
Vitamin PP/B3	DJB05: Vitamin B3 (Total Niacin) EN-HPLC; EN 15652:2009, LC-FLD.

K—Potassium; Ca—Calcium; Mg—Magnesium, P—Phosphorus; Fe—Iron; Cu—Copper; Zn—Zinc; Cl—Chlorine; Mn—Manganese; Se—Selenium; MUFA—Monounsaturated fatty acids; PUFA—Polyunsaturated fatty acids; SFA—Saturated fatty acid; NSP—Non-starch polysaccharide; α—alpha; β—beta; γ—gamma and δ-delta.

3. Results and Discussion

The new data represents the energy, macronutrients (Tables 3–8), mineral (Tables 9 and 10) and vitamin (Tables 11–13) composition per 100 g edible portion of Caribbean, North and West African dishes, snacks and beverages popularly consumed in the UK.

Table 3. Energy, protein, carbohydrate and moisture composition of Caribbean dishes, snacks and beverages in the UK (per 100 g edible portion).

Dish/Snack/Beverage	Moisture (g)	Total Nitrogen (g)	Protein (g)	Carbo-Hydrate (g)	Energy		Starch (g)	Total Sugars (g)	Individual Sugars (g)				
					Kcal	kJ			Gluc	Fruct	Sucr	Malt	Lact
1. Rice and peas	**65.7**	**0.79**	**4.9**	**24.8**	**131.0**	556.0	24.1	0.7	0.1	0.1	0.6	<0.1	<0.1
2. Ackee and saltfish	66.6	1.58	9.9	3.5	203.0	842.0	<0.1	3.5	0.9	1.1	1.5	<0.1	<0.1
3. West Indian Soup	79.1	0.86	5.4	9.9	87.0	364.0	7.6	2.2	0.5	0.4	1.1	0.3	<0.1
4. Goat Curry	66.6	2.99	18.7	<0.1	172.0	717.0	<0.1	1.9	0.5	0.5	0.6	0.2	<0.1
5. Jerk chicken	60.3	4.39	27.4	<0.1	195.0	817.0	<0.1	2.2	0.2	0.3	1.1	0.4	0.3
6. Caribbean fish curry	76.9	1.65	10.3	3.6	100.0	420.0	<0.1	3.6	1.0	1.1	1.2	0.3	<0.1
7. Caribbean vegetable curry	84.8	0.32	2.0	7.4	65.0	271.0	2.4	5.0	1.4	1.3	2.2	<0.1	<0.1
8. Callaloo and Saltfish	72.5	1.32	8.3	4.3	141.0	585.0	<0.1	4.3	1.1	1.1	2.1	<0.1	<0.1
9. Cornmeal porridge	69.3	0.62	3.9	24.0	125.0	528.0	13.3	10.8	<0.1	<0.1	6.7	<0.1	4.1
10. Guinness Punch	81.4	0.50	3.1	15.8	106.0	446.0	<0.1	15.8	0.2	<0.1	11.7	<0.1	3.9
11. Rum punch	73.6	0.01	0.1	26.1	106.0	450.0	11.2	14.9	5.2	5.0	4.7	<0.1	<0.1
12. Fried Dumplings	31.1	0.93	5.8	47.3	318.0	1336.0	46.0	1.3	0.0	0.0	0.2	0.9	0.2
13. Saltfish fritters	49.0	1.51	9.4	24.3	261.0	1093.0	21.9	2.4	0.5	0.4	0.4	0.8	0.4
14. Meat patties	33.9	1.90	11.9	37.4	311.0	1304.0	34.1	3.3	0.4	0.4	0.8	1.7	<0.1

Table 4. Energy, protein, carbohydrate and moisture composition of West African dishes, snacks and beverages in the UK (per 100 g edible portion).

Dish/Snack/Beverage	Moisture (g)	Total Nitrogen (g)	Protein (g)	Carbo-Hydrate (g)	Energy		Starch (g)	Total Sugars (g)	Individual Sugars (g)				
					Kcal	kJ			Gluc	Fruct	Sucr	Malt	Lact
15. Kenkey	**69.3**	**0.42**	**2.6**	**22.5**	**118.0**	498.0	22.2	0.3	0.3	<0.1	<0.1	<0.1	<0.1
16. Shito sauce	17.6	0.79	4.9	14.2	619.0	2555.0	11.2	3.0	0.7	1.3	1.0	<0.1	<0.1
17. Cassava and Plantain fufu	77.4	0.22	1.4	18.9	86.0	366.0	18.8	0.2	<0.1	<0.1	0.2	<0.1	<0.1
18. Malt/Malta drink	84.8	0.06	0.4	14.7	60.0	257.0	3.2	11.6	0.9	0.4	6.5	3.7	<0.1
19. Plantain chips (ripe and chill)	4.0	0.27	1.7	62.1	484.0	2024.0	51.7	10.4	0.7	1.0	8.7	<0.1	<0.1
20. Eba (Gari)	71.0	0.03	0.2	26.3	112.0	474.0	26.0	0.2	0.2	<0.1	<0.1	<0.1	<0.1
21. Rice and Peas/Beans	68.6	0.71	4.4	20.5	118.0	498.0	19.9	0.6	0.1	0.1	0.4	<0.1	<0.1
22. Jollof Rice	66.0	2.50	1.6	23.9	155.0	650.0	22.2	1.7	0.7	0.7	0.3	<0.1	<0.1
23. Egushi Stew	63.1	2.54	15.9	2.4	218.0	904.0	<0.1	2.4	0.7	0.6	0.7	0.4	<0.1
24. Groundnut Soup	73.1	1.69	10.6	3.3	144.0	597.0	1.0	2.3	0.5	0.6	1.1	0.1	<0.1
25. Meat stew	71.3	1.41	8.8	5.0	165.0	684.0	<0.1	5.0	2.1	2.1	0.6	0.2	<0.1
26. One pot pepper soup	78.1	2.24	14.0	1.9	93.0	389.0	<0.1	1.9	0.5	0.6	0.6	0.2	<0.1
27. Okro soup/stew	78.5	2.28	7.3	1.8	120.0	498.0	<0.1	1.8	0.6	0.6	0.5	0.1	<0.1
28. Ewedu soup	75.8	1.91	11.9	1.6	112.0	468.0	<0.1	1.6	0.5	0.5	0.3	0.2	<0.1

Table 5. Energy, protein, carbohydrate and moisture composition of North African dishes, snacks and beverages in the UK (per 100 g edible portion).

Dish/Snack/Beverage	Moisture (g)	Total Nitrogen (g)	Protein (g)	Carbo-Hydrate (g)	Energy		Starch (g)	Total Sugars (g)	Individual Sugars (g)				
					Kcal	kJ			Gluc	Fruct	Sucr	Malt	Lact
29. Chicken Couscous	74.8	1.13	7.1	11.0	**115.0**	480.0	8.3	2.6	0.8	0.7	0.8	0.4	<0.1
30. Vegetable Couscous	68.0	0.68	4.3	18.6	137.0	577.0	12.8	5.9	2.7	2.8	<0.1	0.4	<0.1
31. Lamb Couscous	67.1	1.08	6.8	18.8	141.0	594.0	16.5	2.3	0.5	0.5	1.1	0.3	<0.1
32. Traditional Libyan Soup	78.4	1.29	7.9	1.7	135.0	560.0	<0.1	1.7	0.6	0.7	0.2	0.2	<0.1
33. Kunafa	27.9	0.46	2.9	42.3	413.0	1721.0	22.5	19.8	<0.1	<0.1	18.6	0.5	0.8

Table 6. Fibre and lipid composition of Caribbean dishes, snacks and beverages in the UK (per 100 g edible portion).

Dish/Snack/Beverage	NSP (g)	AOAC Fibre (g)	Fat (g)	SFA (g)	MUFA (g)	PUFA (g)	Trans Fatty Acids (g)
1. Rice and peas	1.9	2.9	0.8	0.3	0.2	0.1	<0.1
2. Ackee and saltfish	1.8	2.3	16.1	4.0	9.0	2.4	<0.1
3. West Indian Soup	1.3	2.0	2.4	1.0	0.9	0.4	<0.1
4. Goat Curry	1.5	2.0	10.4	3.1	5.2	1.4	0.2
5. Jerk chicken	—	—	9.5	2.3	4.5	2.3	<0.1
6. Caribbean fish curry	1.9	3.0	4.3	1.8	1.6	0.7	<0.1
7. Caribbean vegetable curry	2.1	1.1	2.8	0.7	1.0	0.9	<0.1
8. Callaloo and Saltfish	2.2	3.0	9.4	1.7	5.1	2.2	<0.1
9. Cornmeal porridge	0.5	0.9	1.3	0.8	0.3	<0.1	<0.1
10. Guinness Punch	—	—	3.4	2.2	0.8	0.1	<0.1
11. Rum punch	0.3	0.1	0.1	<0.1	<0.1	<0.1	<0.1
12. Fried Dumplings	2.7	2.6	11.2	1.2	6.4	3.0	<0.1
13. Saltfish fritters	2.5	2.2	13.5	1.3	8.0	3.6	<0.1
14. Meat patties	2.3	3.0	12.0	4.0	5.4	2.0	<0.1

__: Food not analysed for the selected nutrient because it was not considered to be an important source of the nutrient.

Table 7. Fibre and lipid composition of West African dishes, snacks and beverages in the UK (per 100 g edible portion).

Dish/Snack/Beverage	NSP (g)	AOAC Fibre (g)	Fat (g)	SFA (g)	MUFA (g)	PUFA (g)	Trans Fatty Acids (g)
15. Kenkey	2.2	3.3	1.2	0.2	0.4	0.5	<0.1
16. Shito sauce	—	—	60.3	19.8	27.1	10.5	0.2
17. Cassava and Plantain fufu	0.8	1.4	0.2	0.1	0.1	<0.1	<0.1
18. Malt/Malta drink	—	—	<0.1	<0.1	<0.1	<0.1	<0.1
19. Plantain chips (ripe and chill)	2.1	5.5	24.2	10.8	10.9	1.3	0.1
20. Eba (Gari)	1.1	2.0	0.2	<0.1	0.1	<0.1	<0.1
21. Rice and Peas/Beans	1.3	4.5	1.0	0.2	0.4	0.4	<0.1
22. Jollof Rice	0.8	1.7	5.5	0.8	2.8	1.7	<0.1
23. Egushi Stew	1.5	2.6	15.5	7.2	5.7	1.9	<0.1
24. Groundnut Soup	1.6	2.1	9.3	2.1	6.1	0.7	<0.1
25. Meat stew	1.7	2.4	11.6	1.5	7.1	2.5	<0.1
26. One pot pepper soup	0.7	1.2	3.0	1.3	1.3	0.2	<0.1
27. Okro soup/stew	1.5	2.7	8.7	3.9	3.5	0.9	<0.1
28. Ewedu soup	2.3	2.3	5.9	2.4	2.4	0.8	<0.1

__: Food not analysed for the selected nutrient because it was not considered to be an important source of the nutrient.

Table 8. Fibre and lipid composition of North African dishes, snacks and beverages in the UK (per 100 g edible portion).

Dish/Snack/Beverage	NSP (g)	AOAC Fibre (g)	Fat (g)	SFA (g)	MUFA (g)	PUFA (g)	Trans Fatty Acids (g)
29. Chicken Couscous	1.5	2.0	4.3	1.2	1.5	1.4	<0.1
30. Vegetable Couscous	2.3	3.6	4.3	1.5	1.4	1.2	<0.1
31. Lamb Couscous	2.1	2.7	3.7	1.1	1.2	1.1	<0.1
32. Traditional Libyan Soup	1.0	1.2	10.4	4.5	3.5	1.2	0.8
33. Kunafa	0.9	1.0	25.5	13.2	7.0	3.6	0.6

Table 9. Inorganic constituents of Caribbean dishes, snacks and beverages in the UK (per 100 g edible portion).

Dish/Snack/Beverage	Na (mg)	K (mg)	Ca (mg)	Mg (mg)	P (mg)	Fe (mg)	Cu (mg)	Zn (mg)	Cl (mg)	Mn (mg)	Se (µg)
1. Rice and peas	15	15	61	0.4	0.2	0.3	345	0.2	<370	15	15
2. Ackee and saltfish	51	28	51	0.8	0.2	0.3	1280	0.2	<370	51	28
3. West Indian soup	32	15	65	0.5	<0.1	0.4	470	0.1	<370	32	15
4. Goat curry	42	28	160	3.0	0.2	2.8	1000	0.2	<370	42	28
5. Jerk chicken	120	32	261	1.4	0.1	2.2	1430	0.1	<370	120	32
6. Caribbean fish curry	35	24	93	1.0	<0.1	<0.2	1250	0.2	<370	35	24
7. Caribbean vegetable curry	33	20	267	0.9	0.1	<0.2	780	0.3	<370	33	20
8. Callaloo and saltfish	115	32	40	1.2	<0.1	0.2	1250	0.3	<370	115	32
9. Cornmeal porridge	97	14	93	<0.4	<0.1	0.2	100	<0.1	<370	97	14
10. Guinness punch	124	13	96	0.6	<0.1	0.2	110	0.1	<370	124	13
11. Rum punch	6	4	3	0.4	<0.1	<0.2	10	0.3	<370	6	4
12. Fried dumplings	256	14	329	1.3	<0.1	0.3	380	0.4	<370	256	14
13. Saltfish fritters	92	15	85	0.9	0.1	0.5	620	0.3	<370	92	15
14. Meat patties	78	21	113	1.9	0.1	2.0	740	0.4	<370	78	21

Table 10. Inorganic constituents of African dishes, snacks and beverages in UK (per 100 g edible portion).

Dish/Snack/Beverage	Na (mg)	K (mg)	Ca (mg)	Mg (mg)	P (mg)	Fe (mg)	Cu (mg)	Zn (mg)	Cl (mg)	Mn (mg)	Se (μg)
West Africa											
15. Kenkey	11	29	72	2.2	<0.1	0.9	580	0.2	<370	11	29
16. Shito sauce	63	35	81	3.6	0.2	0.7	1210	0.9	<370	63	35
17. Cassava and plantain fufu	11	12	32	<0.4	<0.1	<0.2	30	0.1	<370	11	12
18. Malt/Malta drink	3	6	16	<0.4	<0.1	<0.2	20	<0.1	<370	3	6
19. Plantain chips (ripe and chill)	13	94	84	1.2	0.3	0.5	560	0.5	<370	13	94
20. Eba (Gari)	20	1	13	0.6	<0.1	<0.2	70	0.1	<370	20	1
21. Rice and peas/beans	16	24	71	0.8	0.1	0.6	315	0.3	<370	16	24
22. Jollof Rice	15	11	48	0.8	0.1	0.4	620	0.2	<370	15	11
23. Egushi Stew	313	64	306	3.6	0.3	2.3	920	0.7	<370	313	64
24. Groundnut soup	26	41	118	1.6	0.2	1.8	540	0.3	<370	26	41
25. Meat stew	24	20	79	1.7	0.1	2.1	790	0.2	<370	24	20
26. One pot pepper soup	21	15	87	2.2	0.1	2.2	880	0.1	<370	21	15
27. Okro soup/stew	51	29	82	1.8	0.1	1.2	820	0.1	<370	51	29
28. Ewedu soup	76	27	102	3.6	0.2	1.9	900	0.5	<370	76	27
North Africa											
29. Chicken couscous	20	20	82	0.5	0.1	0.5	260	0.2	<370	20	20
30. Vegetable couscous	33	25	83	0.8	0.2	0.5	230	0.5	<370	33	25
31. Lamb couscous	30	32	114	1.1	0.2	1.2	340	0.6	<370	30	32
32. Traditional Libyan soup	24	15	68	0.9	0.1	1.1	340	0.2	<370	24	15
33. Kunafa	53	9	49	0.4	<0.1	0.2	80	0.2	<370	53	9

Table 11. Vitamin composition of Caribbean dishes, snacks and beverages in the UK (per 100 g edible portion).

Sample Dish/Snack/Beverage	α-Tocopherol (mg)	β-Tocopherol (mg)	Biotin (μg)	Calcium Pantothenate (mg)	δ-Tocopherol (mg)	Folic Acid Total (μg)	γ-Tocopherol (mg)	β-Carotene (μg)	Sum Tocopherols (mg)	Vit A (μg)
1. Rice and peas	0	0	1.61	0.154	0	11.4	0	0	0	NA
2. Ackee and saltfish	1.85	0	4.41	0.0816	0	23	1.23	92.2	3.08	0
3. West Indian soup	0.513	0	3.52	0.309	0	16.1	0	310	0.513	0
4. Goat curry	0.734	0	4.94	0.325	0	11.8	0.627	74.2	1.36	0
5. Jerk chicken	1.82	0	12.5	1.25	0	22.2	0	78.4	1.82	0
6. Caribbean fish curry	1.58	0	3.54	0.15	0	17.9	0	2740	1.58	0
7. Caribbean vegetable curry	0.923	0	4.43	0.278	0	20.9	0	629	0.923	NA
8. Callaloo and saltfish	2.4	0	4.75	0.0764	0	16.7	2.24	612	4.64	0
9. Cornmeal porridge	0	0	3.03	0.389	0	7.11	0	15.6	0	NA
10. Guinness punch	0	0	1.91	0.438	0	38.3	0	47	0	NA
11. Rum punch	0	0	0	0.0151	0	8.01	0	111	0	NA
12. Fried dumplings	1.03	0	1.59	0.197	0	9.59	1.88	0	2.91	NA
13. Saltfish fritters	1.85	0	2.41	0.223	0	24.6	2.25	0	4.1	0
14. Meat patties	0.767	0	2.64	0.308	0	22.9	0.879	11.8	1.65	24.3

Sample Dish/Snack/Beverage	Vit B1 (mg)	Vit B12 (μg)	Vit B2 (Riboflavin) (mg)	Vit B5 (Panthotenic Acid) (mg)	Vit B6 (Pyridoxine) (mg)	Vit C (mg)	Vit D2 (μg)	Vit PP/B3, (mg)
1. Rice and peas	0.064	0	0	0.142	0	0	NA	0.24
2. Ackee and saltfish	0.018	1.01	0.0436	0.0751	0.0692	2.62	0	0.446
3. West Indian soup	0.139	0	0.0261	0.284	0.0558	2.26	0	1.08
4. Goat curry	0.032	2.13	0.0678	0.299	0.0683	0	0	2.83
5. Jerk chicken	0.094	0.634	0.112	1.15	0.082	NA	0	5.69
6. Caribbean fish curry	0.03	0.833	0.0284	0.138	0.073	5.76	0	1.04
7. Caribbean vegetable curry	0.028	0	0.0512	0.256	0.0662	0	NA	0.397
8. Callaloo and saltfish	0.015	0.884	0.0453	0.0703	0.0844	0	0	0.362
9. Cornmeal porridge	0.027	0.376	0.137	0.358	0	0	NA	0.134
10. Guinness punch	0.085	0.629	0.295	0.403	0.121	NA	NA	1.02
11. Rum punch	0	0	0	0.0139	0	0.829	NA	0.335
12. Fried dumplings	0.181	0	0	0.181	0	NA	NA	0.854
13. Saltfish fritters	0.119	0.817	0	0.205	0.0513	NA	0	0.563
14. Meat patties	0.201	0.729	0.0482	0.283	0.0688	0	0	1.66

NA—Not Analysed.

Table 12. Vitamins composition of West African dishes, snacks and beverages in the UK (per 100 g edible portion).

Sample Dish/Snack/Beverage	α-Tocopherol (mg)	β-Tocopherol (mg)	Biotin (µg)	Calcium Pantothenate (mg)	δ-Tocopherol (mg)	Folic Acid Total (µg)	γ-Tocopherol (mg)	β-Carotene (µg)	Sum Tocopherols (mg)	Vit A (µg)
15. Kenkey	0.092	0	2.92	0.049	0	11.7	0	0	0.092	NA
16. Shito sauce	13.8	0.577	5.2	0.172	2.43	9.22	10.8	78.7	27.7	0
17. Cassava and plantain fufu	0	0	0	0.199	0	6.63	0	0	0	NA
18. Malt/Malta drink	0.358	0	2.97	0.755	0	8.45	0	0	0.358	NA
19. Plantain chips	3.01	0	9.87	0.29	0	36.4	0	84	3.01	0
20. Eba (Gari)	0	0	0.124	0	7.68	0	0	0	0	NA
21. Rice and peas/beans	0	0	2.16	0.26	0.69	31	0.703	5.71	1.39	NA
22. Jollof rice	0.843	0	1.94	0.212	0	34.9	0.658	76	1.5	0
23. Egushi stew	1.55	0	5.51	0.273	0	33.6	2.37	897	3.92	0
24. Groundnut soup	1.89	0	16.1	0.214	0	35.4	1.18	359	3.07	0
25. Meat stew	2.51	0	6.22	0.128	0	15.9	2.23	132	4.74	0
26. One pot pepper soup	0.3	0	2.24	0.0887	0	8.67	0	41.2	0.3	0
27. Okro soup/stew	1.39	0	4.46	0.181	0	25.4	0	705	1.39	0
28. Ewedu soup	1.76	0	6.56	0.272	0	43.2	0	717	1.76	0

Sample Dish/Snack/Beverage	Vit B1 (mg)	Vit B12 (µg)	Vit B2 (Riboflavin) (mg)	Vit B5 (Panthotenic Acid) (mg)	Vit B6 (Pyridoxine) (mg)	Vit C (mg)	Vit D2 (µg)	Vit PP/B3, (mg)
15. Kenkey	0.097	0	0.0466	0.0451	0.138	0	NA	0.622
16. Shito sauce	0.032	0	0.131	0.159	0.121	NA	0	1.74
17. Cassava and plantain fufu	0	0	0	0.183	0.0664	0	NA	0.319
18. Malt/malta drink	0.297	0	0.0593	0.695	0.231	0	NA	1.58
19. Plantain chips	0.024	0	0.0399	0.267	0.144	NA	NA	0.724
20. Eba (Gari)	0	0	0	0.114	0	0	NA	0.191
21. Rice and peas/ beans	0.055	0	0	0.239	0	NA	NA	0.483
22. Jollof rice	0.043	0	0	0.195	0.0421	0	0	0.747
23. Egushi stew	0.02	0.97	0.0283	0.251	0.0463	0	0	1.3
24. Groundnut soup	0.039	0.938	0	0.197	0.0715	0.518	0	2.9
25. Meat stew	0.034	0.803	0	0.118	0.0847	2.82	0	1.79
26. One pot pepper soup	0.028	1.64	0.0221	0.0816	0.0563	0	0	1.52
27. Okro soup/stew	0.036	0.962	0.0268	0.166	0.0595	0	0	1.33
28. Ewedu soup	0.024	1.25	0.0624	0.251	0.107	2.35	0	1.6

NA—Not Analysed.

Table 13. Vitamins composition of North African dishes, snacks and beverages in the UK (per 100 g edible portion).

Sample Dish/Snack/Beverage	α-Tocopherol (mg)	β-Tocopherol (mg)	Biotin (µg)	Calcium Pantothenate (mg)	δ-Tocopherol (mg)	Folic Acid Total (µg)	γ-Tocopherol (mg)	β-Carotene (µg)	Sum Tocopherols (mg)	Vit A (µg)
29. Chicken couscous	1.05	0	2.37	0.36	0	18.8	1.31	1446	2.37	0
30. Vegetable couscous	0.708	0	3.31	0.191	0	24.3	1.75	119	2.46	NA
31. Lamb couscous	0.54	0	3.18	0.248	0	18	1.41	660	1.95	0
32. Traditional Libyan soup	0.558	0	3.69	0.135	0	12.1	0.746	98.8	1.3	0
33. Kunafa	1.13	0	1.48	0.119	0	7.13	3.18	0	4.31	200

Sample Dish/Snack/Beverage	Vit B1 mg	Vit B12 µg	Vit B2 (Riboflavin) (mg)	Vit B5 (Panthotenic Acid) (mg)	Vit B6 (Pyridoxine) (mg)	Vit C (mg)	Vit D2 (µg)	Vit PP/B3, (mg)		
29. Chicken couscous	0.046	0	0	0.331	0.0714	5.72	0	2.01		
30. Vegetable couscous	0.063	0.492	0	0.176	0.0544	0.568	NT	0.798		
31. Lamb couscous	0.084	0.569	0	0.228	0.0779	1.57	0	0.782		
32. Traditional Libyan soup	0.029	1.09	0.016	0.124	0.0467	1.83	0	1.51		
33. Kunafa	0.032	0	0	0.11	0	0	0	0.387		

NA—Not Analysed.

3.1. Moisture, Energy, Carbohydrate, Protein and Fat Composition

All the foods analysed contained moisture ranging from 4 to 84.8 g/100 g (Tables 3–5). The wide variation in the moisture content is attributed to the type of ingredients and cooking method used. *Shito* sauce and plantain chips require deep fat frying which results in a decrease in moisture with a simultaneous increase in oil [32], hence the low moisture content of these two foods.

Calculated energy values (Tables 3–5) ranged from 60 kcal in Malta drink to 619 kcal (per 100 g edible portion) in the *shito* sauce. For the *shito* sauce, the ingredient of the highest amount is oil, hence the high energy value recorded. An observational study by Goff et al. [21] reported that in the UK the principal sources of energy in the adult Caribbean diet included *rice and peas* and sugar sweetened beverages, whereas for Ghanaians it was *jollof* rice. These foods are however lower in energy than *shito* sauce in the current study. These new food composition data would allow for better quantification of nutrient intake and recommendation of serving size in these population groups. It would also enable health care professionals to identify which foods to encourage or otherwise, when providing dietary advice. Carbohydrate level ranged from less than 0.1 g (in *jerk* chicken and goat curry) to 62.1 g/100 g edible portion of plantain chips (Tables 3–5).

The relationship between dietary carbohydrate intake and risk of hypertension, stroke, type 2 diabetes and obesity, all of which are predominant in people of African and Caribbean ethnicities in the UK [1,2,4] continue to receive a lot of attention. Recently there has been specific focus on carbohydrates and type 2 diabetes. US academics and clinicians are calling for carbohydrate restricted diets as a first approach to prevention and management of type two diabetes [33]. The British Dietetics Association now advise supporting people's choice of low carbohydrate diets for weight loss and diabetes management [34]. On the other hand, the Scientific Advisory Committee on Nutrition [35] considered evidence from both prospective cohort studies and randomised controlled trials on carbohydrates and health. The committee concluded that total carbohydrate intake appears to be neither detrimental nor beneficial to cardio-metabolic (including cardiovascular disease, insulin resistance, glycaemic response and obesity) health. The main starch containing foods were fried dumplings, salt fish, meat patties, *keneky*, *fufu*, plantain chips, *eba*, *rice and peas*, *jollof* rice and *kunafa* (Tables 3–5). The review by SACN [35] reported no association between total starch intake and incidence of coronary events or type 2 diabetes. Corn porridge, *kunafa* and the sugar-sweetened beverages (Malt/Malta, Guinness and rum punch) contained the highest amounts of total sugars. Sucrose levels were mostly less than 1 g but highest in *kunafa* (18.6 g) and Guinness punch (11.7 g). Lactose levels were general less than 0.1 g/100 g of edible portion, therefore negligible (Tables 3–5). However, high consumption of sugar-sweetened beverages is associated with type 2 diabetes and weight gain in children and teenagers [36,37].

In addition, a review by SACN [35] indicated that limited intake of free sugars (total of Non Milk Extrinsic Sugars and added sugars) could reduce the risk of heart disease, type 2 diabetes, bowel health and tooth decay hence the recommendation to limit intakes to 19 g or 5 sugar cubes for children aged 4 to 6, 24 g or 6 sugar cubes for children aged 7 to 10 and 30 g or 7 sugar cubes for 11 years and over, based on average population diets. The current food composition data shows that sucrose levels did not exceed the SACN recommendation for both adults and children.

Increased intakes of total dietary fibre, especially cereal fibre and wholegrain are strongly associated with a lower risk of cardio-metabolic disease [35]. Plantain chips contained the highest amount of fibre of 5.5 g/100 g (Tables 6–8). Non-starch polysaccharide (NSP) levels ranged from 0.3 to 23.7 g/100 g of edible portion of food (Tables 6–8). For those who regularly consumed the dishes, snacks and beverages analysed in the current study, other sources of dietary fibre would need to be included in their diet in order to meet the SACN [35] recommendations (fibre intake of 30 g a day for those aged 16 and over, 25 g for 11 to 15-year-olds, 20 g for 5 to 11-year-olds and 15 g for 2 to 5-year-olds).

The protein content (Tables 3–5) of most of the dishes, snacks and beverages apart from rum punch, Malta drink and *eba* was above 1 g with *jerk* chicken containing the highest amount of 27.4 g/100 g. The contributors of protein were from animal, fish and vegetable sources hence the noticeably low levels in the beverages (rum punch and Malta drink) and *eba* which is made from ground cassava.

Although the protein composition of the vegetable dishes (e.g., vegetable couscous, 4.3 g of protein/100 g of edible portion) were comparatively lower, current evidence suggests that dietary patterns based on more plant sources of protein, or that include unprocessed animal protein also low in saturated fats, could reduce the risk of cardiovascular diseases [38]. Thus, these new data could provide guidance on cardiovascular health in both the majority and Black ethnic populations in the UK [39].

Total fat includes triglycerides, phospholipids, sterols and related compounds. Only *shito* sauce, plantain chips and *kunafa* contained over 20 g/100 g of fat (Tables 6–8). *Shito* sauce, plantain chips, *Egushi* stew and *kunafa* contained over 5 g of saturated fatty acids (SFA)/100 g edible portion of food (Tables 6–8). They also contained comparatively high levels of monounsaturated fatty acids, MUFA. Nearly half the samples analysed had less than 1 g of polyunsaturated fatty acids (PUFA) per 100 g of edible portion of food. Furthermore, *trans* fatty acids (TFA) levels were generally less than 1 g per edible portion of food, hence considered negligible. The main fatty acids present in the foods analysed were SFA, MUFA and PUFA. With reference to current nutrition labelling guidance in UK, *shito* sauce, plantain chips, *egushi* stew and *kunafa* would be classified as high fat foods because they contained over 5 g SFA/100 g edible portion of food [40,41]. A key focus of dietary advice and guidelines is the four fatty acids (TFA, SFA, MUFA, n-3 PUFA and n-6 PUFA) because of their reported association with cardiovascular disease risk [42–44]. However, the previous notion that dietary SFAs lead to increase in serum cholesterol and thus contribute to the risk of cardiovascular disease risk [45] has been challenged [46]. A review by Hammad et al. [47] found that replacing SFA and TFA with n-6 PUFA, n-3 PUFA, or MUFA might protect cardiovascular health but the optimal amount of PUFA or MUFA that can be used to replace SFA and TFA was not identified.

3.2. Mineral Composition

Generally, there were wide variations in the mineral content of the dishes, beverages and snacks analysed. This could be attributed to factors such as variations in ingredients, recipes, cooking or processing methods and brands. The most abundant minerals were Na, K, Ca, Cu, Mn and Se, whereas Mg, P, Fe and Zn were present in small amounts (Tables 9 and 10). Generally, chloride level was less than 370 mg per 100 g edible portion of all the dishes, beverages and snacks.

Sodium (Na) levels in the dishes, beverages and snacks ranged from 3 to 313 mg/100 g (1 gram of sodium per 100 g = 2.5 grams salt). High salt intake is strongly linked to raised blood pressure which increases the risk of heart disease and stroke; common and major causes of death in Europe and UK [48,49]. Although salt intake in the UK is currently on a steady downward trend, levels are 8 g per day on average, therefore above the recommendation of no more than 6 grams per person per day for adults. A reduction in average salt intake from 8 g to 6 g per day is estimated to prevent over 8000 premature deaths each year and save the UK National Health Service (NHS) over £570 million annually [50]. A review by Van-Horn [51] concluded that recommendations to reduce sodium intakes to 2400 mg/d were beneficial. Thus, these traditional dishes, beverages and snacks would increase the low salt options for consumers, which could lead to reduction in overall daily salt intake.

There is increasing evidence to suggest that lower potassium intake or serum potassium levels are associated with a higher risk for type 2 diabetes [52–54]. Although potassium levels (Tables 9 and 10) were less than the UK recommendation of 3.5 mg/day for adults [55], intervention studies are needed to prove that high intakes or supplementation can improve glucose metabolism.

There is evidence to suggest lower calcium intake below the lower reference nutrient intake (LRNI) in some UK minority groups especially women of Black and Asian ethnicities and living on low income [56]. There was calcium present in all the dishes, beverages and snacks analysed (Tables 9 and 10). However, to ensure adequate intake, individuals who regularly consume these dishes would need to include other calcium rich foods in their diet. The latest NDNS data shows that mean intakes of vitamin D were below the RNI (reference nutrient intake) in all age/sex groups and therefore at greater risk of developing a deficiency [50]. About 15 minutes daily exposure to sunlight is recommended. Taking a daily supplement of 10 μg vitamin D is also recommended for the UK population, especially ethnic minority groups from African, Afro-Caribbean and South Asian backgrounds with dark skin and /or cover their bodies when outdoors for cultural reasons, who may not get enough exposure to sunlight [50].

In the UK, around 48% of girls 11 to 18 years and women aged 19 to 64 years have iron intakes below the LRNI and with evidence of anaemia [50]. Iron deficiency anaemia has been associated with low offspring birthweight, can increase susceptibility to infection, and also impact on cognitive development of children and adolescents [57,58]. Data from UK dietary surveys including the Low Income Diet and Nutrition Survey (LIDNS) [26,56,59–62] suggest that iron intakes or status in some South Asian and Black African-Caribbean ethnic minority populations is lower than their White British counterparts. However, according to the SACN [63] report on iron and health, available data suggest that iron intakes of minority ethnic groups aged 16 years and over are not below those of the general UK population. The lack of reliable data on biochemical markers of iron status in UK Black population would account for differences in reported iron intakes and status. The iron content of the dishes, beverages and snacks in the current study were low and ranged from <0.2 to 2.8 mg/100 g edible portion of food (Tables 9 and 10). Thus, individuals who regularly consume these dishes, beverages and snacks will need to include other sources of iron in their diet to prevent the risk of anaemia.

The zinc content was generally very low (Tables 9 and 10) and therefore these foods are not adequate sources of this micronutrient. Zinc is required for growth and normal function of the immune system. Although Zn deficiency is associated with poor growth and increased risk of infection, there is no reliable biomarker to identify the status of this micronutrient [64,65].

Selenium was present in each dish, beverage and snack although levels were varied with plantain chips containing the highest amount—94 µg/100 g of edible portion. In the UK, a substantial proportion of adults aged 19 years and over have selenium intake below the LRNI but the health implications of this are unclear [50].

Jerk chicken, *callaloo* and saltfish, fried dumplings and *egushi* stew contained higher levels of most of the nutrients but they are high in fat. If adequate portion sizes are consumed, they would provide health benefits especially to these three population groups that have been shown to be vulnerable to inadequate micronutrient intake. It is important to note that the adequacy of micronutrient intakes of individuals who regularly consumed these foods depend on various factors including food preparation method, portion size, frequency of consumption and bioavailability rather than just the mineral content per 100 g of the food. Furthermore, reliable biomarkers are needed for better assessment of micronutrient status.

3.3. Vitamins

Vitamin A (Tables 11–13) was only present in *kunafa* and meat patties (200 µg and 24.3/100 g of food, respectively). However, β-carotene was present in twenty-five of the foods analysed. In addition, results from the NDNS showed that mean daily intake of most vitamins derived from dietary sources were close to or above the RNI [5]. Based on UK recommendation for vitamin A [55], *kunafa* could contribute about a third of the RNI of vitamin A (i.e., representing RNI of about 33% for females, 29% for males (11 years and over); 50% RNI for children age 1 to 10 years). However, this dessert is high in sugar and therefore modified recipe (containing reduced sugar) should be adopted by those who consume it. Vitamin D was not present in any of the foods. People of Black ethnicity are among the groups identified as vulnerable to vitamin D deficiency [50]. It is therefore crucial that, this population group increases their exposure to sunlight and also take supplements to avoid the risk of deficiency since dietary sources are unlikely to meet the current RNI of 10 µg per day [66]. There is growing interest around the bioavailability, metabolism, nonantioxidant activity and the role of the various forms of vitamin E in human diseases. Alpha-and gamma-tocopherols are considered the two major forms of the vitamin depending on the source [67]. European Food Safety Authority, EFSA [68] defined Adequate Intake of alpha-tocopheral as 13 mg/day for men, 11 mg/day for women, 6 mg/day for children aged 1 to <3 years (both sexes), 9 mg/day for children aged 3 to <10 years (both sees), for children aged 10 to <18 years, 13 mg/day for boys and 11 mg/day for girls and for infants aged 7–11 months, this was set at 5 mg/day. *Shito* could contribute adequate amount of alpha-tocopherol to the diet. This is likely to be due to the high PUFA content of this sauce.

The water-soluble vitamin composition of the food also varied. Folate was present in most foods, unlike vitamins C and B12. Folate levels range from 0 to 43.2 μg/100 g which is below the RNI for all age groups. The UK government has launched a consultation in 2019 to consider the practicality and of mandatory folic acid fortification, along with the controls on voluntary fortification [69]. The absence or low levels of vitamin C may be attributable to heat losses during cooking. Groundnut soup and goat curry contained the highest amount of biotin (16.1 μg/100 g) and vitamin B12 (2.13 μg/100 g), respectively. Similarly, the highest concentration of calcium pantothenate (1.25 mg/100 g) and vitamin B5 (1.15 mg/100 g) were found in *jerk* chicken and vitamin B1 (0.297 mg/100 g) in Malta drink. Guinness punch had the most amount of vitamins B2 (0.295 mg/100 g) and B6 (0.231 mg/100 g).

3.4. Comparison with Similar Foods in the UK Nutrient Database (McCance and Widdowson's the Composition of Foods)

The only similar food identified in McCance and Widdowson's The composition of foods [70] was ripe plantain, fried in vegetable oil. This was different in composition to the plantain chips in the current data. For instance, the moisture content was higher (34.7 g vs. 4 g), fat lower (9.2 g vs. 24.2 g) and lower energy (267 kcal vs. 484 g) in the ripe plantain, fried in vegetable oil compared to the plantain chips in the current study. The differences would be due to the cooking method, variety of plantain and degree of ripening. In addition, fried plantain is usually consumed as part of a dish whereas plantain chips are snacks. Furthermore, plantain chips in the current study are prepacked samples that are thinly-sliced and deep fat fried to reduce moisture content and enhance shelf life. Foods such as *rice and peas* and *jerk* chicken are very popularly consumed among both the majority and ethnic minority populations in the UK [19–31], but nutrient data for these foods are not available in the McCance and Widdowson food composition tables [70].

3.5. Strengths and Limitations

The study team comprised of trained researchers (including a food scientist and registered nutritionists) with experience in food composition and analyses, and who are of African or Caribbean ethnicity. As previously described by Apekey et al. [14], the various sources, approaches and interview probing questions used enabled the identification of popularly consumed foods. The use of focus group interviews and 24-h dietary recalls also enabled the identification of foods regularly consumed, determination of the frequency of consumption over a period and also improved precision. The interviews and focus group discussions lasted for an hour and therefore allowed for the researchers to capture detailed information on traditional foods, recipes, cooking methods and frequency of consumption. The use of volunteers of the relevant ethnicities in the food preparation in a university Nutrition kitchen allowed for variations in recipes and cooking methods to be taken into account, thereby enhancing the authenticity of the dishes, beverages and snacks. Furthermore, the food samples were analysed in a UK accredited laboratory with trained staff and rigorous quality assurance procedures were followed to ensure the data obtained is reliable and valid.

Limitations of the study include possible introduction of selection bias by the use of convenient sampling to recruit volunteer. However, this approach of sampling through community partnerships or organisations has been shown to improve recruitment of minority ethnic groups into health-related research [71]. The use of 24-h recall may introduce recall bias since it relies on the memory of the volunteers [72]. Although analysing individual foods instead of composite ones may improve the representativeness of the samples analysed, this approach is more complex, time consuming [73] and beyond the scope of the present study.

3.6. Implications for Future Research and Practice

These new nutrient data will contribute to ongoing interactive educational workshops with local communities, and nutrition education and resources in diabetes clinics. The data will allow for better quantification of nutrient intake and recommendations for appropriate serving sizes in these population groups. Furthermore, they will also enable health care professionals to identify which foods to encourage or otherwise, when providing dietary advice. The data will be made available to international and relevant European agencies, for inclusion in their Nutrient Databanks such as the UK's McCance and Widdowson's The Composition of Foods. It will also be made available to health authorities, policy makers and other bodies that have direct influence on promoting health and wellbeing. The data will have various potential uses including (1) contribution to the evidence base of food habits and diet quality, of direct value to nutritional surveys and health surveillance in African and Caribbean populations in the UK and elsewhere in Europe, (2) provide information for components of health promotion programmes contributing to addressing health inequalities and improving quality of life, (3) provide accurate energy and nutrient composition of key dishes for more reliable nutrition labelling and (4) further contributing to health promotion, and food composition data, and the nutrition, dietetics and public health curriculum in the UK and elsewhere.

Supplementary Materials: The following are available online at http://www.mdpi.com/2304-8158/8/10/500/s1, Figure S1: Stages in the prioritisation of popular dishes, snacks and beverages., Figure S2: Composite samples preparation protocol, Figure S3: Stages in the preparation of composite samples.

Author Contributions: Conceptualization, T.A.A.; Methodology, T.A.A.; Formal analysis, T.A.A., M.J.M. and J.C.; Investigation, T.A.A., J.C., N.H.K., O.A.T., M.K., D.W. and M.J.M.; Resources, T.A.A., M.J.M., J.C., N.H.K., O.A.T.; M.K.; D.W. and.; Data curation, T.A.A., M.J.M., J.C., N.H.K., O.A.T., M.K., D.W. and Writing—original draft preparation, T.A.A.; Writing—T.A.A., M.J.M., J.C., N.H.K., O.A.T., M.K., D.W. and.; Supervision, T.A.A.; Project administration, T.A.A., M.J.M., J.C. and.; Funding acquisition, T.A.A., M.J.M.

Funding: The research was internally funded by Leeds Beckett University Early Career, Research Cluster awards and Higher Education Innovation Fund (HEIF).

Acknowledgments: We are grateful to all our volunteers who gave their time to take part in the focus group discussions, interviews and cooking sessions. Our sincere gratitude also goes to Judy Springer, Denisa Copeland and Joan Fishley for their help with the cooking sessions.

Conflicts of Interest: The authors declare no conflict of interest.

References

1. Balfour, P.C.; Rodriguez, C.J.; Ferdinand, K.C. The role of hypertension in race-ethnic disparities in cardiovascular disease. *Curr. Cardiovasc. Risk Rep.* **2015**, *9*, 18. [CrossRef] [PubMed]

2. Tillin, T.; Hughes, A.D.; Godsland, I.F.; Whincup, P.; Forouhi, N.G.; Welsh, P.; Sattar, N.; McKeigue, P.M.; Chaturvedi, N. Insulin resistance and truncal obesity as important determinants of the greater incidence of diabetes in Indian Asians and African Caribbeans compared with Europeans: The Southall and Brent REvisited (SABRE) cohort. *Diabetes Care* **2013**, *36*, 383–393. [CrossRef] [PubMed]

3. Donin, A.S.; Nightingale, C.M.; Owen, C.G.; Rudnicka, A.R.; McNamara, M.C.; Prynne, C.J.; Stephen, A.M.; Cook, D.G.; Whincup, P.H. Ethnic differences in blood lipids and dietary intake between UK children of black African, black Caribbean, South Asian, and white European origin: The Child Heart and Health Study in England (CHASE). *Am. J. Clin. Nutr.* **2010**, *92*, 776–783. [CrossRef] [PubMed]

4. Harding, S.; Whitrow, M.; Lenguerrand, E.; Maynard, M.; Teyhan, A.; Cruickshank, J.K.; Der, G. Emergence of ethnic differences in blood pressure in adolescence. The Determinants of Adolescent Social well-Being and Health study. *Hypertension* **2010**, *55*, 1063–1069. [CrossRef]

5. Public Health England, PHE. National Diet and Nutrition Survey. Results from Years 7 and 8 (Combined) of the Rolling Programme (2014/2015 to 2015/2016). 2018. Available online: https://assets.publishing.service.gov.uk/government/uploads/system/uploads/attachment_data/file/699241/NDNS_results_years_7_and_8.pdf (accessed on 3 July 2019).

6. Office of National Statics. 2011 Census Analysis: Ethnicity and Religion of the Non-UK Born Population in England and Wales Articles. 2015. Available online:

https://www.ons.gov.uk/peoplepopulationandcommunity/culturalidentity/ethnicity/articles/
2011censusanalysisethnicityandreligionofthenonukbornpopulationinenglandandwales/2015-06-18
(accessed on 11 July 2019).

7. Office for National Statistics. Migration Statistics Quarterly Report. 2017. Available online: https://www.ons.gov.uk/peoplepopulationandcommunity/populationandmigration/internationalmigration/bulletins/migrationstatisticsquarterlyreport/may2017 (accessed on 11 July 2019).

8. Medical Research Council, MRC. National Diet and Nutrition Survey. 2018. Available online: https://www.mrc-ewl.cam.ac.uk/research/nutrition-surveys-and-studies/national-diet-and-nutrition-survey/ (accessed on 20 July 2019).

9. Sproston, K.; Mindell, J. Health Survey for England 2004: The Health of Minority Ethnic Groups. 2005. Available online: https://digital.nhs.uk/data-and-information/publications/statistical/health-survey-for-england/health-survey-for-england-2004-health-of-ethnic-minorities-headline-results#key-facts (accessed on 22 July 2019).

10. Erens, B.; Primatesta, P.; Prior, G. Health Survey for England—The Health of Minority Ethnic Groups '99. 2001. Available online: http://webarchive.nationalarchives.gov.uk/20131205105355/http://www.archive.official-documents.co.uk/document/doh/survey99/hse99-00.htm (accessed on 22 July 2019).

11. Mintel. Mintel Report on Ethnic Restaurants and Takeaways, UK; 2016. 2017. Available online: http://store.mintel.com/ethnic-restaurants-and-takeaways-uk-february-2016 (accessed on 20 July 2019).

12. Nielsen. World Foods Growth of Category Opportunities for Retailers and Manufacturers. 2014. Available online: http://www.nielsen.com/content/dam/nielsenglobal/eu/nielseninsights/pdfs/World%20Foods_final_12%2003%2014.pdf (accessed on 20 July 2019).

13. Swan, G.; Dodhia, S.; Farron-Wilson, M.; Powell, N.; Bush, M. Food composition data and public health. *Nutr. Bull.* **2015**, *40*, 223–226. [CrossRef]

14. Apekey, T.A.; Copeman, J.; Kime, N.; Tashani, O.; Kittaneh, M.; Walsh, D.; Maynard, M. Methods of producing new nutrient data for popularly consumed multi-ethnic foods in the UK. *J. Food Compos. Anal.* **2019**, *78*, 9–18. [CrossRef]

15. Greenfield, H.; Southgate, D.A.T. *Food Composition Data Production, Management and Use*; Food and Agriculture Organization of the United Nations: Rome, Italy, 2003.

16. Charrondiere, U.R.; Rittenschober, D.; Nowak, V.; Stadlmayr, B.; Wijesinha-Bettoni, R.; Haytowitz, D. Improving food composition data quality: Three new FAO/INFOODS guidelines on conversions, data evaluation and food matching. *Food Chem.* **2016**, *193*, 75–81. [CrossRef]

17. Mintel. Mintel Report on Attitudes Towards World Cuisines, UK. 2017. Available online: https://reports.mintel.com/display/792389/# (accessed on 10 October 2019).

18. Mintel. Mintel Report on World Cuisines, UK-March 2019. 2019. Available online: https://store.mintel.com/world-cuisines-uk-march-2019 (accessed on 18 July 2019).

19. Asante, M.; Pufulete, M.; Thomas, J.; Wiredu, E.; Intiful, F. Food consumption pattern of Ghanaians living in Accra and London. *IJCR* **2015**, *7*, 16216–16223.

20. Gibson, R.; Knight, A.; Asante, M.; Thomas, J.; Goff, L.M. Comparing dietary macronutrient composition and food sources between native and diasporic Ghanaian adults. *Food Nutr. Res.* **2015**, *59*, 27790. [CrossRef]

21. Goff, L.M.; Timbers, L.; Style, H.; Knight, A. Dietary intake in Black British adults; An observational assessment of nutritional composition and the role of traditional foods in UK Caribbean and West African diets. *Public Health Nutr.* **2015**, *18*, 2191–2201. [CrossRef]

22. Leung, G.; Stanner, S. Diets of minority ethnic groups in the UK: Influence on chronic disease risk and implications for prevention. *Nutr. Bull.* **2011**, *36*, 161–198. [CrossRef]

23. Gandy, J. (Ed.) Nutrition in specific groups. In *Manual of Dietetic Practice*, 5th ed.; Wiley-Blackwell: Hoboken, NJ, USA, 2014; pp. 104–128.

24. Earland, J.; Campbell, J.; Srivastava, A. Dietary habits and health status of African-Caribbean adults. *J. Hum. Nutr. Diet.* **2010**, *23*, 264–271. [CrossRef] [PubMed]

25. Gilbert, P.A.; Khokhar, S. Changing dietary habits of ethnic groups in Europe and implications for health. *Nutr. Rev.* **2008**, *66*, 203–215. [CrossRef] [PubMed]

26. Vyas, A.; Greenhalgh, A.; Cade, A.; Sanghera, B.; Riste, L.; Sharma, S.; Cruickshank, K. Nutrient intakes of an adult Pakistani, European and African-Caribbean community in inner city Britain. *J. Hum. Nutr. Diet.* **2003**, *16*, 327–337. [CrossRef] [PubMed]

27. Sharma, S.; Cade, J.; Landman, J.; Cruickshank, J.K. Assessing the diet of the British African-Caribbean population: Frequency of consumption of foods and food portion sizes. *Int. J. Food Sci. Nutr.* **2002**, *53*, 439–444. [CrossRef] [PubMed]

28. Sharma, S.; Cruickshank, J.K. Cultural differences in assessing dietary intake and providing relevant dietary information to British African-Caribbean populations. *J. Hum. Nutr. Diet.* **2001**, *14*, 449–456. [CrossRef] [PubMed]

29. Sharma, S.; Cade, J.; Jackson, M.; Mbanya, J.C.; Chungong, S.; Forrester, T.; Bennett, F.; Wilks, R.; Balkau, B.; Cruickshank, J.K. Development of food frequency questionnaires in three population samples of African origin from Cameroon, Jamaica and Caribbean migrants to the UK. *Eur. J. Clin. Nutr.* **1996**, *50*, 479–486.

30. Sharma, S.; Cade, J.; Riste, L.; Cruickshank, J.K. Nutrient intake trends among African-Caribbeans in Britain: A migrant population and its second generation. *Public Health Nutr.* **1999**, *2*, 469–476. [CrossRef]

31. Scott, P.; Rajan, L. Eating habits and reactions to dietary advice among two generations of Caribbean people: A South London study, part 1. *Pract. Diabetes Int.* **2000**, *17*, 183–186. [CrossRef]

32. Manjunatha, S.S.; Ravi, N.; Negi, P.S.; Raju, P.S.; Bawa, A.S. Kinetics of moisture loss and oil uptake during deep fat frying of Gethi (Dioscorea kamoonensis Kunth) strips. *J. Food Sci. Technol.* **2014**, *51*, 3061–3071. [CrossRef]

33. Feinman, R.D.; PogozelskiPh, W.K.; Astrup, A.; Bernstein, R.K.; Fine, E.J.; Westman, E.C.; Accurso, A.; Frassetto, L.; Gower, B.A.; McFarlane, S.I.; et al. Dietary carbohydrate restriction as the first approach in diabetes management: Critical review and evidence base. *Nutrition* **2015**, *31*, 1–13. [CrossRef] [PubMed]

34. British Dietetic association, BDA. Low carbohydrate diets for the management of Type 2 Diabetes in adults. 2018. Available online: https://www.bda.uk.com/news/view?id=220&x%5B0%5D=%2Fnews%2Flist (accessed on 15 July 2019).

35. Scientific Advisory Committee on Nutrition, SACN. Carbohydrates and Health. 2015. Available online: https://www.gov.uk/government/uploads/system/uploads/attachment_data/file/445503/SACN_Carbohydrates_and_Health.pdf (accessed on 15 July 2019).

36. Imamura, F.; O'Connor, L.; Ye, Z.; Mursu, J.; Hayashino, Y.; Bhupathiraju, S.N.; Forouhi, N.G. Consumption of sugar sweetened beverages, artificially sweetened beverages, and fruit juice and incidence of type 2 diabetes: Systematic review, meta-analysis, and estimation of population attributable fraction. *BMJ* **2015**, *351*, h3576. [CrossRef] [PubMed]

37. Wang, M.; Yu, M.; Fang, L.; Hu, R.Y. Association between sugar-sweetened beverages and type 2 diabetes: A meta-analysis. *J. Diabetes Investig.* **2015**, *6*, 360–366. [CrossRef] [PubMed]

38. Richter, C.K.; Skulas-Ray, A.C.; Champagne, C.M.; Kris-Etherton, P.M. Plant protein and animal proteins: Do they differentially affect cardiovascular disease risk? *Adv. Nutr.* **2015**, *6*, 712–728. [CrossRef]

39. Public Health England. Health Profile for England. 2017. Available online: https://www.gov.uk/government/publications/health-profile-for-england/chapter-2-major-causes-of-death-and-how-they-have-changed (accessed on 10 July 2019).

40. Department of Health. Guide to Creating a Front of Pack (FoP) Nutrition Label for Pre-Packed Products Sold Through Retail Outlets. 2016. Available online: https://www.gov.uk/government/uploads/system/uploads/attachment_data/file/566251/FoP_Nutrition_labelling_UK_guidance.pdf (accessed on 5 July 2019).

41. Food Standards Agency. Food Labelling. Nutrition Claims. Available online: http://labellingtraining.food.gov.uk/module3/overview_3.html (accessed on 21 July 2019).

42. Gillingham, L.G.; Harris-Janz, S.; Jones, P.J. Dietary monounsaturated fatty acids are protective against metabolic syndrome and cardiovascular disease risk factors. *Lipids* **2011**, *46*, 209–228. [CrossRef]

43. Hooper, L.; Martin, N.; Abdelhamid, A.; Davey, S.G. Reduction in saturated fat intake for cardiovascular disease. *Cochrane Database Syst. Rev.* **2015**, *6*, CD011737. [CrossRef]

44. Krishnan, S.; Cooper, J.A. Effect of dietary fatty acid composition on substrate utilization and body weight maintenance in humans. *Eur. J. Nutr.* **2014**, *53*, 691–710. [CrossRef]

45. Keys, A. Coronary heart disease in seven countries. *Circulation* **1970**, *41* (Suppl. I), 186–195. [CrossRef]

46. Lichtenstein, A.H. Dietary trans fatty acids and cardiovascular disease risk: Past and present. *Curr. Atheroscler. Rep.* **2014**, *16*, 433. [CrossRef]

47. Hammad, S.; Pu, S.; Jones, P.J. Current evidence supporting the link between dietary fatty acids and cardiovascular disease. *Lipids* **2016**, *51*, 507–517. [CrossRef]

48. Vaskonen, M.D. Dietary minerals and modification of cardiovascular risk factors. *J. Nutr. Biochem.* **2003**, *14*, 492–506. [CrossRef]

49. He, J.; Whelton, P.K.; Appel, L.J.; Charleston, J.; Klag, M.J. Long-term effects of weight loss and dietary sodium reduction on incidence of hypertension. *Hypertension* **2000**, *35*, 544–549. [CrossRef] [PubMed]

50. Public Health England. National Diet and Nutrition Survey Results from 2008 to 2017 Assessing Time and Income Trends for Diet, Nutrient Intake and Nutritional Status for the UK. 2019. Available online: https://www.gov.uk/government/statistics/ndns-time-trend-and-income-analyses-for-years-1-to-9 (accessed on 3 July 2019).

51. Van Horn, L. Dietary Sodium and Blood Pressure: How Low Should We Go? *Prog. Cardiovasc. Dis.* **2015**, *58*, 61–68. [CrossRef] [PubMed]

52. Carter, P.; Gray, L.J.; Troughton, J.; Khunti, K.; Davies, M.J. Fruit and vegetable intake and incidence of type 2 diabetes mellitus: Systematic review and meta-analysis. *BMJ* **2010**, *341*, c422. [CrossRef]

53. Ekmekcioglu, C.; Elmadfa, I.; Meyer, A.; Moeslinger, T. The role of dietary potassium in hypertension and diabetes. *J. Physiol. Biochem.* **2016**, *72*, 93–106. [CrossRef]

54. Shin, D.; Joh, H.K.; Kim, K.H.; Park, S.M. Benefits of potassium intake on metabolic syndrome: The fourth Korean National Health and Nutrition Examination Survey (KNHANES IV). *Atherosclerosis* **2013**, *230*, 80–85. [CrossRef]

55. Department of Health. *Dietary Reference Values for Food Energy and Nutrients for the United Kingdom*; Report of the Panel on Dietary Reference Values of the Committee on Medical Aspects of Food Policy (COMA). Report on Health and Social Subjects RHSS No. 32; HMSO: London, UK, 1991.

56. Nelson, M.; Erens, B.; Bates, B.; Church, S.; Boshier, T. *Low Income Diet and Nutrition Survey: Summary of Key Findings*; TSO: London, UK, 2007.

57. Alwan, N.A.; Cade, J.E.; McArdle, H.J.; Greenwood, D.C.; Hayes, H.E.; Simpson, N.A. Maternal iron status in early pregnancy and birth outcomes: Insights from the baby's vascular health and iron in pregnancy study. *Br. J. Nutr.* **2015**, *113*, 1985–1992. [CrossRef]

58. European Food Safety Authority, EFSA. Scientific opinion on the substantiation of a health claim related to iron and cognitive development in children pursuant of Article 14 of Regulation (EC) No 1924/2006. *EFSA J.* **2009**, *7*, 1360. [CrossRef]

59. Donin, A.S.; Nightingale, C.M.; Owen, C.G.; Rudnicka, A.R.; McNamara, M.C.; Prynne, C.J.; Stephen, A.M.; Cook, D.G.; Whincup, P.H. Nutritional composition of the diets of South Asian, black African-Caribbean and white European children in the United Kingdom: The child heart and health study in England (CHASE). *Br. J. Nutr.* **2010**, *104*, 276–285. [CrossRef]

60. Falaschetti, E.; Chaudhury, M. Blood analytes. In *Health Survey for England 2004: The Health of Minority Ethnic Groups*; Sproston, K., Mindell, J., Eds.; The NHS Information Centre: Leeds, UK, 2006; pp. 301–344.

61. Rees, G.A.; Doyle, W.; Srivastava, A.; Brooke, Z.M.; Crawford, A.; Costeloe, K.L. The nutrient intakes of mothers of low birth weight babies - a comparison of ethnic groups in East London, UK. *Matern. Child Nutr.* **2005**, *1*, 91–99. [CrossRef]

62. Thane, C.W.; Bates, C.J.; Prentice, A. Risk factors for low iron intake and poor iron status in a national sample of British young people aged 4-18 years. *Public Health Nutr.* **2003**, *6*, 485–496. [CrossRef] [PubMed]

63. Scientific Advisory Committee on Nutrition, SACN. Iron and health. 2010. Available online: http://www.fcrn.org.uk/sites/default/files/sacn_iron_and_health_report.pdf (accessed on 4 July 2019).

64. Livingstone, C. Zinc: Physiology, deficiency, and parenteral nutrition. *Nutr. Clin. Pract.* **2015**, *30*, 371–382. [CrossRef] [PubMed]

65. Lowe, N.M.; Dykes, F.C.; Skinner, A.; Patel, S.; Warthon-medina, M.; Decsi, T.; Fekete, K.; Souverein, O.W.; Dullemeijer, C.; Cavelaars, A.E.; et al. EURRECA-Estimating zinc requirements for deriving dietary reference values. *Crit. Rev. Food Sci. Nutr.* **2013**, *53*, 1110–1123. [CrossRef] [PubMed]

66. Scientific Advisory Committee on Nutrition, SACN. Vitamin D and Health Report. 2016. Available online: https://www.gov.uk/government/publications/sacn-vitamin-d-and-health-report (accessed on 10 July 2019).

67. Rizvi, S.; Raza, S.T.; Ahmed, F.; Ahmad, A.; Abbas, S.; Mahdi, F. The Role of Vitamin E in Human Health and Some Diseases. *Sultan Qaboos Univ. Med. J.* **2014**, *14*, e157–e165.

68. European Food Safety Authority, EFSA. Scientific Opinion on Dietary Reference Values for vitamin E as α-tocopherol. *Efsa J.* **2015**, *13*, 4149.

69. Public Health England. Open Consultation. Adding Folic Acid to Flour. 2019. Available online: https://www.gov.uk/government/consultations/adding-folic-acid-to-flour (accessed on 16 July 2019).

70. Food Standard Agency & Public Health England. *The Composition of Foods*, 7th ed.; Royal Society of Chemistry: London, UK, 2015.

71. Bonevski, B.; Randell, M.; Paul, C.; Chapman, K.; Twyman, L.; Bryant, J.; Brozek, I.; Hughes, C. Reaching the hard-to-reach: A systematic review of strategies for 413 improving health and medical research with socially disadvantaged groups. *BMC Med. Res. Methodol.* **2014**, *14*, 42. [CrossRef]

72. In National and Scottish Research Studies. Briefing Paper Prepared for: Working Group on Monitoring Scottish Dietary Targets Workshop. Available online: https://www.food.gov.uk/sites/default/files/multimedia/pdfs/scotdietassessmethods.pdf (accessed on 10 September 2019).

73. Greenfield, H. Uses and abuses of food composition data. *Food Aust.* **1990**, *42*, S1–S44.

Article

The Protective Effect of Brazilian Propolis against Glycation Stress in Mouse Skeletal Muscle

Tatsuro Egawa [1,2,*], Yoshitaka Ohno [3], Shingo Yokoyama [3], Takumi Yokokawa [1], Satoshi Tsuda [1], Katsumasa Goto [3,4] and Tatsuya Hayashi [1]

[1] Laboratory of Sports and Exercise Medicine, Graduate School of Human and Environmental Studies, Kyoto University, Kyoto 606-8501, Japan; takumi.yokokawa@gmail.com (T.Y.); tsuda.satoshi.55u@st.kyoto-u.ac.jp (S.T.); tatsuya@kuhp.kyoto-u.ac.jp (T.H.)

[2] Laboratory of Health and Exercise Sciences, Graduate School of Human and Environmental Studies, Kyoto University, Kyoto 606-8501, Japan

[3] Laboratory of Physiology, School of Health Sciences, Toyohashi SOZO University, Toyohashi 440-8511, Japan; yohno@sozo.ac.jp (Y.O.); s-yokoyama@sozo.ac.jp (S.Y.); gotok@sepia.ocn.ne.jp (K.G.)

[4] Department of Physiology, Graduate School of Health Sciences, Toyohashi SOZO University, Toyohashi 440-8511, Japan

* Correspondence: egawa.tatsuro.4u@kyoto-u.ac.jp; Tel.: +81-75-753-6613; Fax: +81-75-753-6885

Received: 20 August 2019; Accepted: 24 September 2019; Published: 25 September 2019

Abstract: We investigated the protective effect of Brazilian propolis, a natural resinous substance produced by honeybees, against glycation stress in mouse skeletal muscles. Mice were divided into four groups: (1) Normal diet + drinking water, (2) Brazilian propolis (0.1%)-containing diet + drinking water, (3) normal diet + methylglyoxal (MGO) (0.1%)-containing drinking water, and (4) Brazilian propolis (0.1%)-containing diet + MGO (0.1%)-containing drinking water. MGO treatment for 20 weeks reduced the weight of the extensor digitorum longus (EDL) muscle and tended to be in the soleus muscle. Ingestion of Brazilian propolis showed no effect on this change in EDL muscles but tended to increase the weight of the soleus muscles regardless of MGO treatment. In EDL muscles, Brazilian propolis ingestion suppressed the accumulation of MGO-derived advanced glycation end products (AGEs) in MGO-treated mice. The activity of glyoxalase 1 was not affected by MGO, but was enhanced by Brazilian propolis in EDL muscles. MGO treatment increased mRNA expression of inflammation-related molecules, interleukin (IL)-1β, IL-6, and toll-like receptor 4 (TLR4). Brazilian propolis ingestion suppressed these increases. MGO and/or propolis exerted no effect on the accumulation of AGEs, glyoxalase 1 activity, and inflammatory responses in soleus muscles. These results suggest that Brazilian propolis exerts a protective effect against glycation stress by inhibiting the accumulation of AGEs, promoting MGO detoxification, and reducing proinflammatory responses in the skeletal muscle. However, these anti-glycation effects does not lead to prevent glycation-induced muscle mass reduction.

Keywords: advanced glycation end products; anti-glycation; glycative stress; glyoxalase; methylglyoxal; cytokine

---

1. Introduction

The skeletal muscle is the largest organ that contributes to maintaining physical locomotive function. It is also a major site of glucose and lipid metabolism and an endocrine organ with myokine secretions [1]. A number of epidemiological studies revealed that people with type 2 diabetes tend to have lower muscle strength and mass [2]. The potential underlying mechanism of this skeletal muscle dysfunction is linked to hyperglycemia, chronic inflammation, and oxidative stress [2].

Glycation is a biochemical process through which reducing sugars like glucose react and bond non-enzymatically with proteins. Glycation stress, which is caused by glycation and includes the formation of advanced glycation end products (AGEs) and a subsequent dysfunction of proteins and/or cellular signaling [3], are considered related with the progress of muscle dysfunctions. It has been reported that elevated AGEs in the blood or skin are negatively correlated with muscle mass, grip strength, and glucose tolerance in the elderly [4–8] and patients with diabetes [9]. Our recent study demonstrated that AGEs suppressed formation of myotubes in C2C12 skeletal muscle cells by deteriorating cellular signal transduction of protein synthesis and suggested that AGEs inhibited skeletal muscle formation and maturation [10]. Furthermore, serum AGE levels are related to diabetic complications in children and young adults with type 1 diabetes [11–13], thus indicating that glycation stress might affect skeletal muscle function regardless of age. In fact, our previous study revealed that the consumption of an AGE-rich diet for 16 weeks in young mice led to degenerative changes in skeletal muscle, including low muscle mass, low grip strength, low force relative to muscle mass, and muscle fatigability [14]. Furthermore, AGEs treatment in skeletal muscle has been illustrated to induce insulin resistance in young male and female rodents [15,16]. Therefore, inhibiting glycation stress is considered an effective strategy for preventing skeletal muscle dysfunction regardless of age.

AGEs lead to the activation of different signaling pathways mediated by several cell surface receptors. The activation of receptors for AGEs (RAGE) is considered as a major mediator of AGE pathogenicity [17,18]. Although the recruitment of RAGE stimulates myogenesis that is important for skeletal muscle development, the chronic stimulation of RAGE, due to high concentration of AGEs, causes myopathy through inflammatory responses [19]. In addition to RAGE, toll-like receptor 4 (TLR4) is involved in AGE-mediated inflammatory responses, such as cytokine production [20]. The interaction between AGEs-RAGE leads to activation of intracellular nuclear factor-κ B and subsequently increases the expression of several proinflammatory cytokines, including tumor necrosis factor-α (TNFα) and interleukin (IL)-6 [21]. Furthermore, AGEs stimulates the secretion of IL-6 through RAGE and/or TLR4 in macrophages [22]. A recent study has also demonstrated that AGEs-induced inflammatory responses occur via IL-1β in human placental cells [23]. These proinflammatory cytokines are known factors of muscle wasting [24] and insulin resistance [25], and thus the suppression of AGEs-associated inflammatory responses can be a target of maintaining muscle functions.

Propolis, a natural resinous substance produced by honeybees, is traditionally used in herbal medicine and has recently been suggested to possess several biological properties including anticancer, antioxidant, and anti-inflammatory activities [26]. The wide diversity of plant species used by bees as resin sources for propolis production determines its chemical diversity by region. Among propolis of various production area, Brazilian propolis contains a number of phenolic compounds such as artepillin C, p-coumaric acids, and kaempferide [27,28], and has become a popular health supplement due to its many biological properties [29]. Recent studies have reported that several polyphenol substances exert anti-glycation functions by inhibiting the formation of AGEs, promoting their degradation, and by exerting an antagonizing effect on AGE receptors [30]. This suggests that Brazilian propolis may possess an anti-glycation capacity and contributes to maintaining skeletal muscle functions. Previous studies demonstrated that European poplar type of propolis have anti-glycation activity in vitro [31–33]. However, no reports have investigated the anti-glycation effects of Brazilian propolis and its efficacy in vivo.

In the present study, we aimed to examine the protective effect of Brazilian propolis against glycation stress in the skeletal muscle. To this end, we subjected the skeletal muscles of mice to glycation stress using methylglyoxal (MGO), a precursor of AGEs, for 20 weeks and investigated the effect of Brazilian propolis on alleviation of this stress.

2. Materials and Methods

2.1. Animals and Treatment

Twenty-four male C57BL/6NCr mice (4-weeks-old) were purchased from Shimizu Breeding Laboratories (Kyoto, Japan). The mice were placed in a room maintained at 22–24 °C with a 12:12 h light/dark cycle. After 1 week of adjustment, the mice were randomly divided into four groups ($n = 6$/group): (1) Normal diet (AIN-93G; Oriental Koubo, Tokyo, Japan) + drinking water (N), (2) Brazilian propolis (0.1%)-containing diet + drinking water (PRO), (3) normal diet + MGO (0.1%)-containing drinking water (MGO), and (4) Brazilian propolis (0.1%)-containing diet + MGO (0.1%)-containing drinking water (MGO + PRO). The Brazilian propolis powder of ethanol extracts (LY-009), standardized to contain a minimum of 8.0% artepillin C was obtained from Yamada Bee Company, Inc. (Okayama, Japan). The Brazilian propolis was originated from Baccharis dracunculifolia of Southeast Brazil. The nutritional information of AIN-93G and Brazilian propolis powder of ethanol extracts is listed in Table 1. The doses of propolis and methylglyoxal, and their duration of intake were determined by previous experimental studies [34,35]. For each group, all mice were housed in a single cage and provided free access to food and drinking water for 20 weeks. Body weight was measured once every two weeks. Food and fluid intakes were measured during two consecutive days every two weeks and averaged as grams per day per mouse.

Table 1. Nutritional information of AIN-93G and Brazilian propolis powder of ethanol extracts.

Components	AIN-93G (per 100 g)	Brazilian Propolis Powder (per 100 g)
Carbohydrate	63.0 g	4.2 g
Protein	20.0 g	0.7 g
Fat	7.0 g	47.0 g
Mineral	3.5 g	0.4 g
Vitamin	1.0 g	
Calories	400 kcal	758 kcal

At the end of the study period, the slow-twitch soleus muscle and fast-twitch extensor digitorum longus (EDL) muscles and tibia were collected from each mouse under anesthesia using mixtures of medetomidine hydrochloride (0.3 mg/kg), midazolam (4.0 mg/kg), and butorphanol (5.0 mg/kg). All animal protocols were carried out in accordance with the Guide for the Care and Use of Laboratory Animals by the National Institutes of Health (Bethesda, MD, USA) and were approved by the Kyoto University Graduate School of Human and Environmental Studies (approval number: 28-A-2, approval date: 2016.3.29).

2.2. Anti-Glycation Assay

The anti-glycation activity of propolis was performed using the Albumin Glycation Assay Kit (AAS-AGE-K01, Cosmo Bio, Tokyo, Japan). Briefly, propolis was dissolved in dimethyl sulfoxide at a concentration of 0, 0.1, 1, 10, and 100 mg/mL, and the solutions were incubated with 50 mM glyceraldehyde and bovine serum albumin solutions for 48 h at 37 °C. The fluorescence of AGEs was estimated using a fluorescence microplate reader equipped with a 355 nm excitation filter and 460 nm emission filter. Inhibitory effects of AGE formation were expressed as percent change relative to the value of a solution containing 20 mM aminoguanidine.

2.3. Measurement of MGO-Derived AGE Content

The MGO-derived AGE content in muscles was measured using an OxiSelect Methylglyoxal Competitive ELISA Kit (STA-811, Cell Biolabs, Milpitas, CA, USA) according to the manufacturer's protocol.

2.4. Measurement of Glyoxalase 1 Activity

The activity of glyoxalase 1 in muscles was measured using a Glyoxalase I Activity Assay Kit (Colorimetric) (K591-100, BioVision, San Diego, CA, USA) according to the manufacturer's protocol.

2.5. Real-Time RT-PCR Analysis

A separate set of muscle samples were subjected to RT-PCR analysis, which was performed as previously described [36]. Total RNA was extracted from frozen muscles using the RNeasy Mini Kit (Qiagen, Venlo, Netherlands). RNA was reverse-transcribed into complementary DNA (cDNA) using PrimeScript RT Master Mix (Perfect Real Time) (Takara Bio, Kusatsu, Japan). Synthesized cDNA was subjected to real-time RT-PCR (Step One Real Time System, Applied Biosystems, Carlsbad, CA, USA) using SYBR Premix Ex Taq II (Takara Bio, Kusatsu, Japan) and then analyzed using StepOne Software v2.3 (Applied Biosystems, Foster City, CA, USA). Relative fold change of expression was calculated by the comparative CT method. β-actin and ribosomal protein S18 (Rps18) was used as an internal standard. Primers used were as follows: Interleukin-1β (IL-1β), 5′-TCCAGGATGAGGACATGAGCAC-3′ (forward) and 5′-GAAC GTCACACACCAGCAGGTTA-3′ (reverse); IL-6, 5′-CCACTTCACAAGTCGGAGGCTTA-3′ (forward), and 5′-TGCAAGTGCATC ATCGTTGTTC-3′ (reverse); toll-like receptor 4 (TLR4), 5′-TCCTGTGGACAAGGTCAGCAAC-3′ (forward) and 5′-TTACACTCAGACTCG GCACTTAGCA-3′ (reverse); receptor for AGE (RAGE), 5′-AGCCACTGGAATTGTCGATGAG-3′ (forward), and 5′-GCTGTGAGTTCAGAGGCAGGA-3′ (reverse); β-actin, 5′-CATCCGTAAAGACCTCTATGCCAAC-3′ (forward), and 5′-ATGGAGCCAC CGATCCACA-3′ (reverse); and Rps18, 5′-TTGGTGAGGTCAATGTCTGCTTT-3′ (forward), and 5′-AA GTTTCAGCACATCCTGCGAGT-3′ (reverse).

2.6. Statistics

All values were expressed as means ± SE. For each group of data, normality (the Kolmogorov–Smirnov test) and equal variance tests (Levene's test) were performed and data that were not normally distributed were log-transformed before the analysis of variance (ANOVA). The statistical significance of differences in body weight, food intake, and fluid intake between groups was determined via a repeated-measures ANOVA. The statistical significance of differences in muscle weight, MGO-derived AGEs content, and mRNA expression was analyzed using two-way ANOVA with propolis and MGO as the main factors. In the event of significant main effects and/or interactions, post hoc Tukey-Kramer tests were performed. Differences between groups were considered statistically significant at $p < 0.05$. All statistical analyses were performed using the Ekuseru-Toukei 2012 software (Social Survey Research Information, Tokyo, Japan).

3. Results

3.1. Anti-Glycation Effects of Brazilian Propolis In Vitro

The inhibitory activity of Brazilian propolis against formation of AGEs was evaluated by measurement of fluorescent AGEs formed by glyceraldehyde and bovine serum albumin (Figure 1). Propolis inhibited the formation of fluorescent AGEs (0 mg/mL, 0 ± 4.53%; 0.1 mg/mL, 31.1 ± 2.87%; 1.0 mg/mL, 84.5 ± 2.00%; 10 mg/mL, 118 ± 0.62%; 100 mg/mL, 122 ± 1.79%, means ± SE, $n = 4$/group).

3.2. The Effect of Brazilian Propolis on Body Weight, Food and Fluid Intake, and Muscle Weight

Body and muscle weights and food and fluid intake are presented in Table 2 and Figures S1–S3. Repeated measures ANOVA did not reveal significant differences in the body weights among the groups ($p = 0.070$). Food intake was significantly different among the groups ($p = 0.0004$); specifically, MGO + PRO group had lower food intake than all other groups ($p = 0.007$ vs. N; $p = 0.003$ vs. PRO; $p = 0.001$ vs. MGO). Fluid intake was significantly different among the groups ($p = 0.0004$); in that, the PRO ($p = 0.010$), MGO ($p = 0.043$), and MGO + PRO ($p = 0.0003$) groups had lower fluid intakes than N group. Two-way ANOVA revealed that MGO, but not propolis, had a significant main effect on EDL muscle weight normalized to tibia length (propolis, $p = 0.69$; MGO, $p = 0.039$) (Table 2) and muscle cross sectional area (CSA) (propolis, $p = 0.95$; MGO, $p = 0.042$) (Table S1). No significant main effects were observed for soleus muscle weight normalized to tibia length (propolis, $p = 0.054$; MGO, $p = 0.086$) (Table 2) and muscle CSA (propolis, $p = 0.18$; MGO, $p = 0.13$) (Table S1). However, propolis and MGO showed a large ($\eta^2 = 0.16$) and moderate ($\eta^2 = 0.12$) effect size in soleus muscle mass as calculated using η^2, respectively.

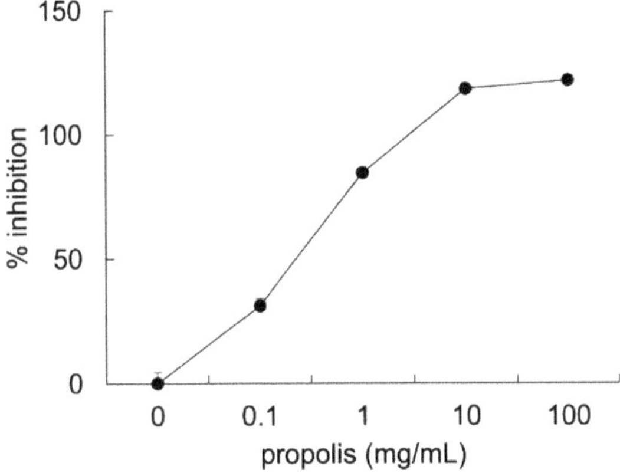

Figure 1. The inhibitory effect of Brazilian propolis used at different concentrations (0, 0.1, 1.0, 10, and 100 mg/mL) on formation of advanced glycation end products (AGEs). Values are means ± SE; $n = 4$/group. Values are expressed as percent change relative to the value of aminoguanidine.

Table 2. Body weight, food intake, fluid intake, and muscle weight.

	Normal	Propolis	MGO	MGO + Propolis	ANOVA
Initial body weight (g)	17.7 ± 0.8	17.7 ± 0.5	17.6 ± 0.5	17.7 ± 0.4	—
Final body weight (g)	41.1 ± 0.7	41.3 ± 0.8	38.5 ± 0.8	40.4 ± 0.5	$p = 0.070$
Food intake (g/day/mouse)	3.8 ± 0.6 [†]	3.7 ± 0.4 [†]	3.7 ± 0.4 [†]	3.4 ± 0.4	$p = 0.0004$
Fluid intake (g/day/mouse)	3.2 ± 0.4	2.8 ± 0.3 *	2.9 ± 0.3 *	2.7 ± 0.3 *	$p = 0.0004$
EDL weight/tibia (mg/mm)	0.65 ± 0.03	0.67 ± 0.02	0.62 ± 0.02	0.62 ± 0.01	Propolis ($p = 0.69$) MGO ($p = 0.039$)
Soleus weight/tibia (mg/mm)	0.55 ± 0.01	0.59 ± 0.01	0.53 ± 0.02	0.56 ± 0.02	Propolis ($p = 0.054$) MGO ($p = 0.086$)

EDL, extensor digitorum longus; MGO, methylglyoxal; $n = 4$–6/group; * and † indicates $p < 0.05$ vs. Normal and MGO + propolis group, respectively.

3.3. Brazilian Propolis Suppressed the Accumulation of MGO-Derived AGEs in the Skeletal Muscle In Vivo

The content of MGO-derived AGEs in EDL and soleus muscles was measured to evaluate the effect of Brazilian propolis on accumulation of AGEs in the skeletal muscle in vivo. In the EDL muscle, two-way ANOVA revealed a significant interaction ($p = 0.020$); specifically, the content of MGO-derived AGEs following MGO treatment tended to increase ($p = 0.08$), but it had a large effect size as calculated using Cohen's d (d = 1.54). Brazilian propolis ingestion suppressed this accumulation ($p = 0.003$) (Figure 2). In the soleus muscle, no significant alterations in the content of MGO-derived AGEs was observed according to ANOVA (MGO, $p = 0.59$; propolis, $p = 0.97$) (Figure 2).

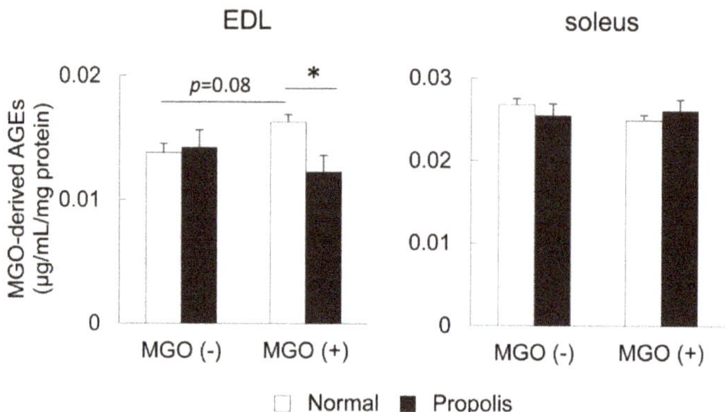

Figure 2. The content of methylglyoxal (MGO)-derived advanced glycation end products (AGEs) in skeletal muscles. The extensor digitorum longus (EDL) and soleus muscles were dissected from mice treated with or without Brazilian propolis (0.1%)-containing diet or MGO (0.1%)-containing drinking water for 20 weeks. Values are means ± SE; n = 5–6/group. * $p < 0.05$ between the groups.

3.4. Brazilian Propolis Enhanced Glyoxalase 1 Activity in The Skeletal Muscle

To evaluate the ability of Brazilian propolis to detoxify MGO in the skeletal muscle, the activity of glyoxalase 1, a dicarbonyl compound eliminating enzyme, was measured in the EDL and soleus muscles. In the EDL, two-way ANOVA revealed a significant main effect of propolis, but not MGO (propolis, $p = 0.038$; MGO, $p = 0.49$) (Figure 3). In the soleus muscle, no significant alterations in glyoxalase 1 activity was observed via ANOVA (MGO, $p = 0.25$; propolis, $p = 0.47$) (Figure 3).

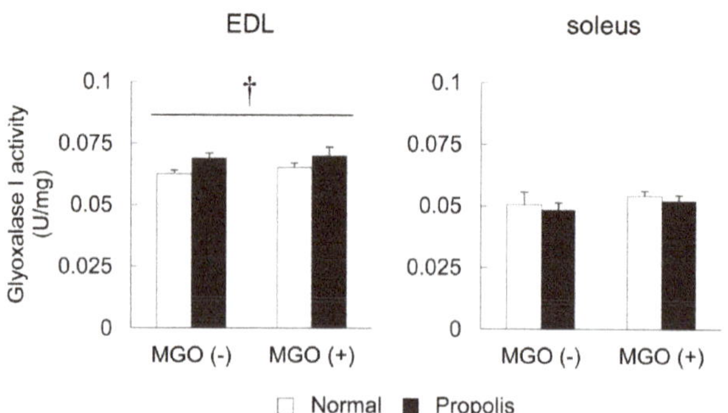

Figure 3. The activity of glyoxalase 1 in skeletal muscles. EDL and soleus muscles were dissected from mice treated with or without Brazilian propolis (0.1%)-containing diet or MGO (0.1%)-containing drinking water for 20 weeks. Values are means ± SE; n = 5–6/group. †, significant main effect between diets (normal and propolis).

3.5. Brazilian Propolis Suppressed MGO-Induced mRNA Expression of Inflammatory-Related Molecules in The Skeletal Muscle

To evaluate the effect of propolis on inflammatory responses, the mRNA expression of proinflammatory cytokines, IL-1β and IL-6, and AGEs-related receptors, TLR4 and RAGE, were measured in the EDL and soleus muscles (Figure 4). In the EDL muscle, two-way ANOVA revealed significant effects on IL-1β, Il-6, and TLR4 expression. MGO treatment significantly increased the mRNA expression of IL-1β ($p = 0.037$); however, the ingestion of Brazilian propolis suppressed this increase ($p = 0.006$). MGO treatment tended to increase the mRNA expression of IL-6, and propolis ingestion suppressed this effect in MGO-treated mice ($p = 0.036$). MGO treatment significantly increased the mRNA expression of TLR4 ($p = 0.028$), but no change was observed under propolis ingestion ($p = 0.20$). The mRNA level of RAGE in the EDL muscle was not altered by either MGO ($p = 0.28$) or propolis ($p = 0.41$). In the soleus muscle, no significant alterations in the mRNA expression of IL-1β (MGO, $p = 0.078$; propolis, $p = 0.44$), IL-6 (MGO, $p = 0.80$; propolis, $p = 0.34$), TLR4 (MGO, $p = 0.21$; propolis, $p = 0.38$), and RAGE (MGO, $p = 0.33$; propolis, $p = 0.16$) were observed via ANOVA (Figure 4).

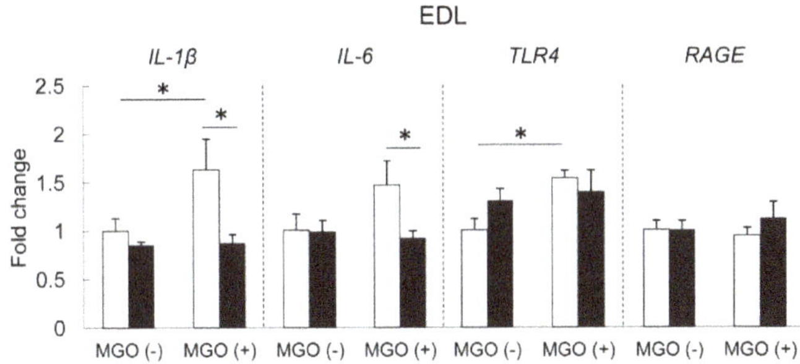

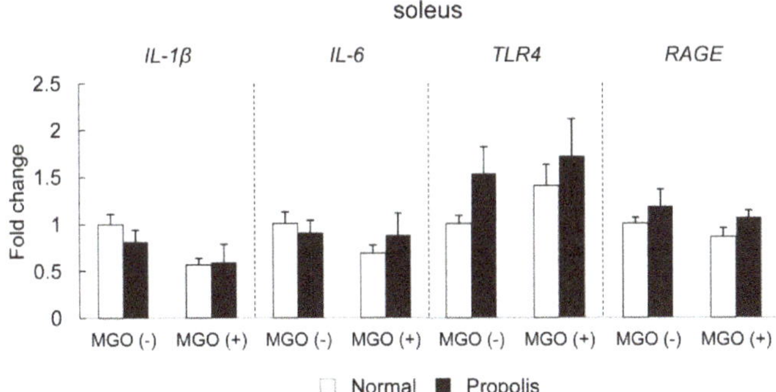

Figure 4. mRNA expression of interleukin (IL)-1β, IL-6, toll-like receptor 4 (TLR4), and receptor for AGEs (RAGE) in skeletal muscles. The EDL and soleus muscles were dissected from mice treated with or without propolis (0.1%)-containing diet or MGO (0.1%)-containing drinking water for 20 weeks. Data of IL-1β in the EDL muscle were log-transformed for normal distribution before analysis of variance (ANOVA). Values are means ± SE; $n = 3$–6/group. * $p < 0.05$ between the groups.

4. Discussion

The current study revealed several novel findings regarding the effect of Brazilian propolis on glycation stress in the skeletal muscle. Firstly, Brazilian propolis inhibited the formation of AGEs in vitro (Figure 1). Secondly, the 20-week ingestion of Brazilian propolis suppressed the accumulation of MGO-derived AGEs (Figure 2), promoted activity of glyoxalase 1 (Figure 3), and attenuated mRNA expressions of proinflammatory cytokines IL-1β and IL-6 (Figure 4) in the EDL but not the soleus muscle.

Glycation stress is suppressed by several mechanisms such as inhibition of AGEs formation, MGO formation, and oxidative stress, detoxification of MGO, and blocked activation of AGEs receptors [30]. To date, many researchers have evaluated the inhibitory effect of natural compounds on the formation of AGEs, and many natural plants are confirmed to reduce glycation stress by inhibiting this formation [37,38]. In this study, we provided evidence for the inhibitory capacity of Brazilian propolis on formation of AGEs in vitro (Figure 1). To the best of our knowledge, this is the first study to demonstrate Brazilian propolis-induced anti-glycation activity. In accordance with this finding, European propolis, which differ from Brazilian propolis in terms of raw materials and components, have been revealed to inhibit glucose-derived and D-ribose-derived AGEs production [31–33]. These findings suggest that various types of propolis have the capacity to inhibit AGEs formation in vitro.

We also provided a subsequent confirmation for the inhibitory effect of Brazilian propolis on formation of AGEs in vivo by showing that Brazilian propolis led to suppression of MGO-derived AGE accumulation in the skeletal muscle of MGO-loaded mice (Figure 2). This protective effect was seen in the fast-type EDL muscle but not the slow-type soleus muscle. Our previous study demonstrated that a 16-week glycation stress induced by a high-AGE diet in mice promoted the accumulation of AGEs in the EDL but not the soleus muscle [14]. Furthermore, another research has shown that the accumulation of AGEs in the diabetic rat skeletal muscle was greater in fast-type muscle [39]. These findings suggest that fast-type muscles are susceptible to AGEs and that Brazilian propolis improves the inhibitory capacity against AGE formation in fast-type muscle. The potential mechanisms regarding the greater susceptibility of fast-type muscles to AGEs have been described. First, slow-type muscles have a higher protein turnover rate than fast-type muscles [40,41], thus indicating that AGEs are more easily broken down in slow-type muscle than fast-type muscle, and fast-type muscles have a tendency to accumulate AGEs. Second, fast-type muscles are more susceptible to changes in nutrients and hormones than slow-type muscles [42], thus indicating that fast-type muscles are more sensitive to AGEs and propolis than slow-type muscles. However, considering the finding that MGO tended to affect muscle mass with a large effect size in soleus muscle (Table 2), additional examinations using other muscles are needed to clear the fiber-type specific susceptibility to glycation stress.

Brazilian propolis increased muscle mass of soleus almost significantly ($p = 0.054$) with a moderate effect size ($\eta^2 = 0.12$), raising a possibility that Brazilian propolis has a hypertrophic effect in soleus muscle regardless of glycation stress. However, there was no significant difference in the calculated muscle CSA (Table S1), indicating that propolis-induced increase in soleus muscle mass was not caused by hypertrophy. In this regard, Brazilian propolis might stimulate glycogen accumulation, and thereby led to muscle mass gain, because it has been shown that Brazilian propolis stimulated glucose uptake in mouse skeletal muscle [43]. However, a previous study has shown that six-week intake of water extract of Korean propolis did not affect glycogen content in the gastrocnemius muscle of rat [44]. Another possibility is that Brazilian propolis increased connective tissue in muscle because it has been shown that propolis stimulated migration and proliferation of fibroblast cells [45]. At present, however, we have no clear explanation for the mechanism by which Brazilian propolis causes gain of soleus muscle mass without hypertrophy.

Detoxification of MGO is also important for reducing glycation stress. MGO is a highly reactive dicarbonyl compound and the major precursor in the formation of AGEs [46,47]. When MGO production exceeds the detoxification capacity, it can modify arginine residues to form MGO-derived AGEs [47]. The most important MGO detoxification system is the glyoxalase system and glyoxalase 1 functions as a rate-limiting enzyme in this system. Under normal physiological conditions, >99% of MGO is metabolized via the glyoxalase system [48]. In the present study, propolis enhanced glyoxalase 1 activity in the EDL muscle (Figure 3), indicating its capability to detoxify MGO, and thereby in inhibition of MGO-derived AGE production. Therefore, in addition to the inhibitory effect of AGE formation, an enhancement of the glyoxalase system mediated by Brazilian propolis may contribute to the inhibitory effect of accumulation of MGO-derived AGEs in the skeletal muscle.

Inflammation is a crucial contributor toward pathology of diseases implicated in skeletal muscle dysfunction [25,49,50]. Binding of AGEs to AGE receptors including RAGE and TLR4 are potent inducers of inflammatory responses [22]. Inhibition of RAGE and TLR4 effectively reversed the AGE-induced inflammatory signaling [22,51]. In the present study, Brazilian propolis showed no effect of mRNA expression of RAGE, but prevented MGO-treated induction of IL-1β, IL-6, and TLR4 (Figure 4). Consistent with this observation, previous studies have shown that propolis inhibits production of IL-1β in human immune cells [52] and IL-6 in murine macrophages [53]. The current study is the first study that shows that Brazilian propolis has a protective effect on AGE-induced inflammatory responses in the skeletal muscle.

Among the various components of Brazilian propolis [27,28], kaempferide [54], ferulic acid [55], and caffeic acid derivatives [56] are established inhibitors of AGE formation. Furthermore, it has been shown that propolis-induced anti-inflammatory responses may occur due to the synergistic effect of its compounds, artepillin C [57], coumaric acid and cinnamic acid [53], and hesperidin, quercetin, and caffeic acid derivatives [52]. Flavonoid compounds also have a stimulating effect on the glyoxalase system and thereby contribute to neuroprotection [58]. Collectively, the protective activity of propolis against glycation stress in the skeletal muscle may be attributed to the combined biological activity of these phenolic compounds.

Food and fluid intakes were significantly affected by propolis and/or MGO treatment (Table 2). Food intake was reduced in the MGO + PRO group compared with that in the other groups, thus suggesting that MGO + PRO group received a lower contribution from propolis. However, the beneficial effects of propolis, including reduced AGEs accumulation and inflammatory responses, were confirmed in this group. Fluid intake was affected by treatment with MGO and/or propolis, but there was no difference between the MGO and MGO + PRO groups, thus indicating that the beneficial effects of propolis in the MGO + PRO group, including reduced MGO-derived AGEs content and inflammatory responses, were not caused by decreased MGO consumption. Therefore, we believe that the difference of food and fluid intakes does not influence the conclusions of this study.

5. Conclusions

The present study revealed that Brazilian propolis protects against MGO-induced glycation stress in mouse skeletal muscles. Brazilian propolis inhibits the accumulation of AGEs, promotes MGO detoxification, and reduces the levels of proinflammatory cytokines. However, Brazilian propolis does not prevent glycation-induced muscle mass reduction. These bioactivities of Brazilian propolis may be effective to protect skeletal muscle dysfunctions induced by aging and pathogenesis.

Supplementary Materials: The following are available online at http://www.mdpi.com/2304-8158/8/10/439/s1, Figure S1: Changes in body weight, Figure S2: Changes in food intake, Figure S1: Changes in fluid intake, Table S1: Muscle cross sectional area.

Author Contributions: Conceptualization, T.E. and S.Y.; Investigation, T.E., Y.O., and S.Y.; Formal analysis, T.E. and Y.O.; Writing—original draft preparation, T.E., K.G., and T.H; Writing—review, S.T., T.Y., and T.H.; Supervision, S.T. and T.Y.; Funding acquisition, T.E, Y.O., S.Y., K.G., and T.H.

Funding: This study was supported in part by Yamada Research Grant (T.E.) and JSPS KAKENHI (T.E., 18H03148 and 19K22806; Y.O., 18K10796; S.Y., 16K16450 and 19K19854; K.G., 17K01762, 18H03160 and 19K22825; T.H., 19K11520); the Council for Science, Technology and Innovation; SIP (Funding agency: Bio-oriented Technology Research Advancement Institution, NARO) (T.H., 14533567). Additional research grants were provided by the Takeda Research Support (T.H., TKDS20170531015), the Science Research Promotion Fund from the Promotion and Mutual Aid Corporation for Private Schools of Japan; and Graduate School of Health Sciences, Toyohashi SOZO University (K.G.).

Conflicts of Interest: The authors declare no conflict of interest.

References

1. Argiles, J.M.; Campos, N.; Lopez-Pedrosa, J.M.; Rueda, R.; Rodriguez-Manas, L. Skeletal muscle regulates metabolism via interorgan crosstalk: Roles in health and disease. *J. Am. Med. Dir. Assoc.* **2016**, *17*, 789–796. [CrossRef] [PubMed]
2. Bianchi, L.; Volpato, S. Muscle dysfunction in type 2 diabetes: A major threat to patient's mobility and independence. *Acta Diabetol.* **2016**, *53*, 879–889. [CrossRef]
3. Lin, J.A.; Wu, C.H.; Lu, C.C.; Hsia, S.M.; Yen, G.C. Glycative stress from advanced glycation end products (ages) and dicarbonyls: An emerging biological factor in cancer onset and progression. *Mol. Nutr. Food Res.* **2016**, *60*, 1850–1864. [CrossRef] [PubMed]
4. Dalal, M.; Ferrucci, L.; Sun, K.; Beck, J.; Fried, L.P.; Semba, R.D. Elevated serum advanced glycation end products and poor grip strength in older community-dwelling women. *J. Gerontol. A Biol. Sci. Med. Sci.* **2009**, *64*, 132–137. [CrossRef] [PubMed]
5. Semba, R.D.; Bandinelli, S.; Sun, K.; Guralnik, J.M.; Ferrucci, L. Relationship of an advanced glycation end product, plasma carboxymethyl-lysine, with slow walking speed in older adults: The inchianti study. *Eur. J. Appl. Physiol.* **2010**, *108*, 191–195. [CrossRef] [PubMed]
6. Momma, H.; Niu, K.; Kobayashi, Y.; Guan, L.; Sato, M.; Guo, H.; Chujo, M.; Otomo, A.; Yufei, C.; Tadaura, H.; et al. Skin advanced glycation end product accumulation and muscle strength among adult men. *Eur. J. Appl. Physiol.* **2011**, *111*, 1545–1552. [CrossRef] [PubMed]
7. Kato, M.; Kubo, A.; Sugioka, Y.; Mitsui, R.; Fukuhara, N.; Nihei, F.; Takeda, Y. Relationship between advanced glycation end-product accumulation and low skeletal muscle mass in japanese men and women. *Geriatr. Gerontol. Int.* **2017**, *17*, 785–790. [CrossRef] [PubMed]
8. Tan, K.C.; Shiu, S.W.; Wong, Y.; Tam, X. Serum advanced glycation end products (ages) are associated with insulin resistance. *Diabetes Metab. Res. Rev.* **2011**, *27*, 488–492. [CrossRef] [PubMed]
9. Mori, H.; Kuroda, A.; Ishizu, M.; Ohishi, M.; Takashi, Y.; Otsuka, Y.; Taniguchi, S.; Tamaki, M.; Kurahashi, K.; Yoshida, S.; et al. Association of accumulated advanced glycation end-products with a high prevalence of sarcopenia and dynapenia in patients with type 2 diabetes. *J. Diabetes Investig.* **2019**, *10*, 1332–1340. [CrossRef] [PubMed]
10. Egawa, T.; Ohno, Y.; Yokoyama, S.; Goto, A.; Ito, R.; Hayashi, T.; Goto, K. The effect of advanced glycation end products on cellular signaling molecules in skeletal muscle. *J. Phys. Fit. Sports Med.* **2018**, *7*, 229–238. [CrossRef]
11. Brunvand, L.; Heier, M.; Brunborg, C.; Hanssen, K.F.; Fugelseth, D.; Stensaeth, K.H.; Dahl-Jorgensen, K.; Margeirsdottir, H.D. Advanced glycation end products in children with type 1 diabetes and early reduced diastolic heart function. *BMC Cardiovasc. Disord.* **2017**, *17*, 133. [CrossRef] [PubMed]
12. Chiarelli, F.; de Martino, M.; Mezzetti, A.; Catino, M.; Morgese, G.; Cuccurullo, F.; Verrotti, A. Advanced glycation end products in children and adolescents with diabetes: Relation to glycemic control and early microvascular complications. *J. Pediatr.* **1999**, *134*, 486–491. [CrossRef]
13. Chiarelli, F.; Catino, M.; Tumini, S.; Cipollone, F.; Mezzetti, A.; Vanelli, M.; Verrotti, A. Advanced glycation end products in adolescents and young adults with diabetic angiopathy. *Pediatr. Nephrol.* **2000**, *14*, 841–846. [CrossRef] [PubMed]
14. Egawa, T.; Tsuda, S.; Goto, A.; Ohno, Y.; Yokoyama, S.; Goto, K.; Hayashi, T. Potential involvement of dietary advanced glycation end products in impairment of skeletal muscle growth and muscle contractile function in mice. *Br. J. Nutr.* **2017**, *117*, 21–29. [CrossRef] [PubMed]
15. Cassese, A.; Esposito, I.; Fiory, F.; Barbagallo, A.P.; Paturzo, F.; Mirra, P.; Ulianich, L.; Giacco, F.; Iadicicco, C.; Lombardi, A.; et al. In skeletal muscle advanced glycation end products (ages) inhibit insulin action and induce the formation of multimolecular complexes including the receptor for ages. *J. Biol. Chem.* **2008**, *283*, 36088–36099. [CrossRef] [PubMed]
16. Pinto-Junior, D.C.; Silva, K.S.; Michalani, M.L.; Yonamine, C.Y.; Esteves, J.V.; Fabre, N.T.; Thieme, K.; Catanozi, S.; Okamoto, M.M.; Seraphim, P.M.; et al. Advanced glycation end products-induced insulin resistance involves repression of skeletal muscle glut4 expression. *Sci. Rep.* **2018**, *8*, 8109. [CrossRef] [PubMed]

17. Ramasamy, R.; Yan, S.F.; Schmidt, A.M. Receptor for age (rage): Signaling mechanisms in the pathogenesis of diabetes and its complications. *Ann. N. Y. Acad. Sci.* **2011**, *1243*, 88–102. [CrossRef] [PubMed]

18. Chuah, Y.K.; Basir, R.; Talib, H.; Tie, T.H.; Nordin, N. Receptor for advanced glycation end products and its involvement in inflammatory diseases. *Int. J. Inflam.* **2013**, *2013*. [CrossRef] [PubMed]

19. Riuzzi, F.; Sorci, G.; Sagheddu, R.; Chiappalupi, S.; Salvadori, L.; Donato, R. Rage in the pathophysiology of skeletal muscle. *J. Cachexia Sarcopenia Muscle* **2018**, *9*, 1213–1234. [CrossRef] [PubMed]

20. Ibrahim, Z.A.; Armour, C.L.; Phipps, S.; Sukkar, M.B. Rage and tlrs: Relatives, friends or neighbours? *Mol. Immunol.* **2013**, *56*, 739–744. [CrossRef]

21. Liu, Y.; Liang, C.; Liu, X.; Liao, B.; Pan, X.; Ren, Y.; Fan, M.; Li, M.; He, Z.; Wu, J.; et al. Ages increased migration and inflammatory responses of adventitial fibroblasts via rage, mapk and nf-kappab pathways. *Atherosclerosis* **2010**, *208*, 34–42. [CrossRef] [PubMed]

22. Ohtsu, A.; Shibutani, Y.; Seno, K.; Iwata, H.; Kuwayama, T.; Shirasuna, K. Advanced glycation end products and lipopolysaccharides stimulate interleukin-6 secretion via the rage/tlr4-nf-kappab-ros pathways and resveratrol attenuates these inflammatory responses in mouse macrophages. *Exp. Ther. Med.* **2017**, *14*, 4363–4370. [PubMed]

23. Seno, K.; Sase, S.; Ozeki, A.; Takahashi, H.; Ohkuchi, A.; Suzuki, H.; Matsubara, S.; Iwata, H.; Kuwayama, T.; Shirasuna, K. Advanced glycation end products regulate interleukin-1beta production in human placenta. *J. Reprod. Dev.* **2017**, *63*, 401–408. [CrossRef] [PubMed]

24. Costamagna, D.; Costelli, P.; Sampaolesi, M.; Penna, F. Role of inflammation in muscle homeostasis and myogenesis. *Mediat. Inflamm.* **2015**, *2015*. [CrossRef] [PubMed]

25. Wei, Y.; Chen, K.; Whaley-Connell, A.T.; Stump, C.S.; Ibdah, J.A.; Sowers, J.R. Skeletal muscle insulin resistance: Role of inflammatory cytokines and reactive oxygen species. *Am. J. Physiol. Regul. Integr. Comp. Physiol.* **2008**, *294*, R673–R680. [CrossRef] [PubMed]

26. Pasupuleti, V.R.; Sammugam, L.; Ramesh, N.; Gan, S.H. Honey, propolis, and royal jelly: A comprehensive review of their biological actions and health benefits. *Oxid. Med. Cell. Longev.* **2017**, *2017*. [CrossRef] [PubMed]

27. Marcucci, M.C.; Ferreres, F.; Custodio, A.R.; Ferreira, M.M.; Bankova, V.S.; Garcia-Viguera, C.; Bretz, W.A. Evaluation of phenolic compounds in brazilian propolis from different geographic regions. *Z. Naturforsch. C* **2000**, *55*, 76–81. [CrossRef]

28. Righi, A.A.; Negri, G.; Salatino, A. Comparative chemistry of propolis from eight brazilian localities. *Evid. Based Complement. Alternat. Med.* **2013**, *2013*. [CrossRef]

29. Kocot, J.; Kielczykowska, M.; Luchowska-Kocot, D.; Kurzepa, J.; Musik, I. Antioxidant potential of propolis, bee pollen, and royal jelly: Possible medical application. *Oxid. Med. Cell. Longev.* **2018**, *2018*. [CrossRef]

30. Yeh, W.J.; Hsia, S.M.; Lee, W.H.; Wu, C.H. Polyphenols with antiglycation activity and mechanisms of action: A review of recent findings. *J. Food Drug. Anal.* **2017**, *25*, 84–92. [CrossRef]

31. Boisard, S.; Le Ray, A.M.; Gatto, J.; Aumond, M.C.; Blanchard, P.; Derbre, S.; Flurin, C.; Richomme, P. Chemical composition, antioxidant and anti-ages activities of a french poplar type propolis. *J. Agric. Food Chem.* **2014**, *62*, 1344–1351. [CrossRef] [PubMed]

32. Sahebi, U.; Divsalar, A. Synergistic and inhibitory effects of propolis and aspirin on structural changes of human hemoglobin resulting from glycation: An in vitro study. *J. Iran. Chem. Soc.* **2016**, *13*, 2001–2011. [CrossRef]

33. Boisard, S.; Shahali, Y.; Aumond, M.-C.; Derbré, S.; Blanchard, P.; Dadar, M.; Le Ray, A.-M.; Richomme, P. Anti-age activity of poplar-type propolis: Mechanism of action of main phenolic compounds. *International J. Food Sci. Technol.* **2019**. [CrossRef]

34. Aoi, W.; Hosogi, S.; Niisato, N.; Yokoyama, N.; Hayata, H.; Miyazaki, H.; Kusuzaki, K.; Fukuda, T.; Fukui, M.; Nakamura, N.; et al. Improvement of insulin resistance, blood pressure and interstitial ph in early developmental stage of insulin resistance in oletf rats by intake of propolis extracts. *Biochem. Biophys. Res. Commun.* **2013**, *432*, 650–653. [CrossRef] [PubMed]

35. Zhao, Y.; Wang, P.; Chen, H.; Sang, S. Dietary quercetin inhibits methylglyoxal-induced advanced glycation end products formation in mice. *FASEB J.* **2016**, *30*, 692.

36. Goto, A.; Ohno, Y.; Ikuta, A.; Suzuki, M.; Ohira, T.; Egawa, T.; Sugiura, T.; Yoshioka, T.; Ohira, Y.; Goto, K. Up-regulation of adiponectin expression in antigravitational soleus muscle in response to unloading followed by reloading, and functional overloading in mice. *PLoS ONE* **2013**, *8*, e81929. [CrossRef] [PubMed]

37. Ishioka, Y.; Yagi, M.; Ogura, M.; Yonei, Y. Polyphenol content of various vegetables: Relationship to antiglycation activity. *Glycative Stress Res.* **2015**, *2*, 41–51.

38. Chinchansure, A.A.; Korwar, A.M.; Kulkarni, M.J.; Joshi, S.P. Recent development of plant products with anti-glycation activity: A review. *RSC Adv.* **2015**, *5*, 31113–31138. [CrossRef]

39. Snow, L.M.; Fugere, N.A.; Thompson, L.V. Advanced glycation end-product accumulation and associated protein modification in type ii skeletal muscle with aging. *J. Gerontol. Ser. A Biol. Sci. Med. Sci.* **2007**, *62*, 1204–1210. [CrossRef]

40. van Wessel, T.; de Haan, A.; van der Laarse, W.J.; Jaspers, R.T. The muscle fiber type-fiber size paradox: Hypertrophy or oxidative metabolism? *Eur. J. Appl. Physiol.* **2010**, *110*, 665–694. [CrossRef]

41. Lewis, S.E.; Kelly, F.J.; Goldspink, D.F. Pre- and post-natal growth and protein turnover in smooth muscle, heart and slow- and fast-twitch skeletal muscles of the rat. *Biochem. J.* **1984**, *217*, 517–526. [CrossRef] [PubMed]

42. Baillie, A.G.; Garlick, P.J. Responses of protein synthesis in different skeletal muscles to fasting and insulin in rats. *Am. J. Physiol.* **1991**, *260*, E891–E896. [CrossRef] [PubMed]

43. Ueda, M.; Hayashibara, K.; Ashida, H. Propolis extract promotes translocation of glucose transporter 4 and glucose uptake through both pi3k- and ampk-dependent pathways in skeletal muscle. *Biofactors* **2013**, *39*, 457–466. [CrossRef] [PubMed]

44. Kwon, T.D.; Lee, M.W.; Kim, K.H. The effect of exercise training and water extract from propolis intake on the antioxidant enzymes activity of skeletal muscle and liver in rat. *J. Exerc. Nutr. Biochem.* **2014**, *18*, 9–17. [CrossRef] [PubMed]

45. Jacob, A.; Parolia, A.; Pau, A.; Davamani Amalraj, F. The effects of malaysian propolis and brazilian red propolis on connective tissue fibroblasts in the wound healing process. *BMC Complement. Altern. Med.* **2015**, *15*, 294. [CrossRef] [PubMed]

46. Ramasamy, R.; Yan, S.F.; Schmidt, A.M. Methylglyoxal comes of age. *Cell* **2006**, *124*, 258–260. [CrossRef] [PubMed]

47. Wetzels, S.; Wouters, K.; Schalkwijk, C.G.; Vanmierlo, T.; Hendriks, J.J. Methylglyoxal-derived advanced glycation endproducts in multiple sclerosis. *Int. J. Mol. Sci.* **2017**, *18*, 421. [CrossRef] [PubMed]

48. Rabbani, N.; Thornalley, P.J. Dicarbonyl proteome and genome damage in metabolic and vascular disease. *Biochem. Soc. Trans.* **2014**, *42*, 425–432. [CrossRef] [PubMed]

49. Londhe, P.; Guttridge, D.C. Inflammation induced loss of skeletal muscle. *Bone* **2015**, *80*, 131–142. [CrossRef] [PubMed]

50. Dalle, S.; Rossmeislova, L.; Koppo, K. The role of inflammation in age-related sarcopenia. *Front. Physiol.* **2017**, *8*, 1045. [CrossRef]

51. Chen, Y.J.; Sheu, M.L.; Tsai, K.S.; Yang, R.S.; Liu, S.H. Advanced glycation end products induce peroxisome proliferator-activated receptor gamma down-regulation-related inflammatory signals in human chondrocytes via toll-like receptor-4 and receptor for advanced glycation end products. *PLoS ONE* **2013**, *8*, e66611.

52. Ansorge, S.; Reinhold, D.; Lendeckel, U. Propolis and some of its constituents down-regulate DNA synthesis and inflammatory cytokine production but induce tgf-beta1 production of human immune cells. *Z. Naturforsch. C* **2003**, *58*, 580–589. [CrossRef]

53. Bachiega, T.F.; Orsatti, C.L.; Pagliarone, A.C.; Sforcin, J.M. The effects of propolis and its isolated compounds on cytokine production by murine macrophages. *Phytother. Res.* **2012**, *26*, 1308–1313. [CrossRef] [PubMed]

54. Yang, R.; Wang, W.X.; Chen, H.J.; He, Z.C.; Jia, A.Q. The inhibition of advanced glycation end-products by five fractions and three main flavonoids from camellia nitidissima chi flowers. *J. Food Drug. Anal.* **2018**, *26*, 252–259. [CrossRef] [PubMed]

55. Silvan, J.M.; Assar, S.H.; Srey, C.; Dolores Del Castillo, M.; Ames, J.M. Control of the maillard reaction by ferulic acid. *Food Chem.* **2011**, *128*, 208–213. [CrossRef] [PubMed]

56. Sasaki, K.; Chiba, S.; Yoshizaki, F. Effect of natural flavonoids, stilbenes and caffeic acid oligomers on protein glycation. *Biomed. Rep.* **2014**, *2*, 628–632. [CrossRef] [PubMed]

57. Paulino, N.; Abreu, S.R.; Uto, Y.; Koyama, D.; Nagasawa, H.; Hori, H.; Dirsch, V.M.; Vollmar, A.M.; Scremin, A.; Bretz, W.A. Anti-inflammatory effects of a bioavailable compound, artepillin c, in brazilian propolis. *Eur. J. Pharmacol.* **2008**, *587*, 296–301. [CrossRef] [PubMed]

58. Frandsen, J.R.; Narayanasamy, P. Neuroprotection through flavonoid: Enhancement of the glyoxalase pathway. *Redox Biol.* **2018**, *14*, 465–473. [CrossRef] [PubMed]

Article

Postprandial Glycemic and Insulinemic Effects of the Addition of Aqueous Extracts of Dried Corn Silk, Cumin Seed Powder or Tamarind Pulp, in Two Forms, Consumed with High Glycemic Index Rice

Sumanto Haldar [1,*], Linda Gan [1], Shia Lyn Tay [1], Shalini Ponnalagu [1] and Christiani Jeyakumar Henry [1,2]

[1] Clinical Nutrition Research Centre (CNRC), Singapore Institute for Clinical Sciences (SICS), Agency for Science Technology and Research (A*STAR), 30 Medical Drive, Singapore 117609, Singapore; lindaalerts@gmail.com (L.G.); tay_shia_lyn@sics.a-star.edu.sg (S.L.T.); shalini_Ponnalagu@sics.a-star.edu.sg (S.P.); jeya_henry@sics.a-star.edu.sg (C.J.H.)

[2] Department of Biochemistry, National University of Singapore, Singapore 117596, Singapore

* Correspondence: sumanto_haldar@sics.a-star.edu.sg

Received: 21 August 2019; Accepted: 19 September 2019; Published: 24 September 2019

Abstract: Several plant-based traditional ingredients in Asia are anecdotally used for preventing and/or treating type 2 diabetes. We investigated three such widely consumed ingredients, namely corn silk (CS), cumin (CU), and tamarind (TA). The aim of the study was to determine the effects of aqueous extracts of these ingredients consumed either as a drink (D) with high-glycemic-index rice or added to the same amount of rice during cooking (R) on postprandial glycemia (PPG), insulinemia (PPI), and blood pressure (BP), over a 3 h measurement period. Eighteen healthy Chinese men (aged 37.5 ± 12.5 years, BMI 21.8 ± 1.67 kg/m^2) took part in a randomized crossover trial, each completing up to nine sessions. Compared to the control meal (plain rice + plain water), the addition of test extracts in either form did not modulate PPG, PPI, or BP. However, the extracts when added within rice while cooking gave rise to significantly lower PPI than when consumed as a drink ($p < 0.01$). Therefore, the form of consumption of phytochemical-rich ingredients can differentially modulate glucose homeostasis. This study also highlights the need for undertaking randomized controlled clinical trials with traditional foods/components before claims are made on their specific health effects.

Keywords: corn silk; cumin; tamarind; aqueous extracts; form; postprandial glycemia; postprandial insulinemia

---

1. Introduction

Arising from traditional dietary and alternative medicine practices across Asia (e.g., traditional Chinese medicine, Ayurveda) there have been a plethora of food-based components claimed as beneficial in improving the risk of diabetes and other cardiometabolic conditions. Unfortunately, the majority of these traditional practices have been based on historical pretext rather than on solid scientific evidence. In fact, robustly controlled clinical trials have been sparse, and most of the research claims came from in vitro studies and/or in vivo animal studies. In general, these traditional ingredients are all plant-based ingredients, rich in polyphenols and other phytonutrients with a wide range of biological functions including α-amylase and α-glucosidase enzyme inhibition, anti-inflammatory and antioxidant properties, insulin secretagogues, etc. [1–3]. In Asia, over centuries these ingredients have been integrated as part of daily diets of individuals into various culinary preparations and thereby used simultaneously as traditional "medicines" to maintain health and reduce disease risk. Three such commonly used traditional ingredients include corn silk, cumin, and tamarind, and some of the prior evidence for their biological effects related to improving glycemic control and/or type 2 diabetes risk are outlined below.

Corn silk (*Stigma maydis*) is a waste by-product of corn and has been long used in traditional Chinese medicine to treat diabetes and associated conditions [4]. Corn silk is rich in various phytonutrients (e.g., polyphenols, terpenoids, etc.), vitamins, and minerals, as detailed elsewhere [5]. There are several mechanisms through which corn silk is thought to improve glycemic control, which include inhibition of α-amylase and α-glucosidase enzymes [6–8], inhibition of inflammation [9], reduced formation of advanced glycation end products (AGEs), reduced oxidative stress, and the enhancement of glucose-stimulated insulin secretion [6,10,11]. Therefore, corn silk has the potential to improve postprandial glycemic control, although to our best knowledge this has not been tested using a randomized controlled trial (RCT) in normoglycemic, non-diabetic individuals.

Similarly, both cumin and tamarind have been used as traditional food ingredients across various Asian cuisines and are considered to have glycemic-improving properties. For example, it has been previously shown that the consumption of cumin seeds by streptozotocin-induced diabetic rats led to significant reductions in hyperglycemia and glucosuria as well as improvements in renal function [12]. Another study showed reductions in both fasted blood glucose and glycated hemoglobin when alloxan-induced diabetic rats were administered cumin over a 6-week period [13], whereas a separate study showed a reduction in postprandial glycemic responses in rabbits fed cumin [14]. More recently, it has also been demonstrated that the cuminaldehyde found in cumin has significant α-glucosidase as well as aldose reductase inhibition [15]. On the other hand, tamarind pulp extract has been shown to improve glycemic, insulinemic, as well as lipidemic responses upon feeding to high-fat-induced obese rats for 10 weeks [16]. Several other studies using rat/mice models of diabetes have also shown improvements in glycemic parameters with various extracts of tamarind [17–20]. While the collective evidence in animal studies for both cumin and tamarind having the potential to improve glycemic control is rather substantial, there have been hardly any randomized controlled trials in humans.

Therefore, in this "Phytochemical Rich Ingredients as Functional Foods (PRIFF)" study, the primary aims were to assess the three ingredients, viz., corn silk, cumin, and tamarind as listed above, in terms of their effects on postprandial glycemia and insulinemia, when consumed with a carbohydrate-rich meal principally consisting of a high-glycemic-index (GI) white rice. The secondary aim was to also measure the effects of these three ingredients on postprandial blood pressure. The study used aqueous extracts which can be obtained using household preparation methods. For cumin and tamarind, we used dietary doses, whereas for corn silk we used two separate doses—the "high dose" consisting of an amount commonly used in traditional Chinese medicine, and a lower (one-half of the high) dose to assess whether this lower, "dietary" dose would still be effective in improving glycemic parameters. Moreover, there is now considerable evidence that the food form plays an important role in determining glycemic and/or insulinemic responses [21–23]. We therefore investigated all the test ingredients in two separate forms: (i) extracts consumed as a drink (D) with plain rice or (ii) extracts added into rice (R) during cooking and consumed with plain water.

2. Materials and Methods

2.1. Recruitment and Ethics

Eighteen volunteers were enrolled in this study through flyers placed around The National University of Singapore and from our center's recruitment database of volunteers who had previously given consent to be contacted for research. Inclusion criteria for this study were: Chinese, male, age between 21 to 60 years old, body mass index (BMI) between 18.5 and 25.0 kg/m^2, waist circumference $\leq$90 cm, fasting blood glucose <7.0 mmol/L, blood pressure <140 mmHg and <90 mmHg for systolic and diastolic, respectively. Interested volunteers attended a screening visit after an overnight fast (no food or drink except water for a minimum of 10 h). At the screening visit, height was measured using a stadiometer (Seca 217, Hamburg, Germany), weight and body fat percentage were measured using the Tanita analyzer (BC-418, Tokyo, Japan), seated resting blood pressure was measured using an automated sphygmomanometer (Omron HEM907, Singapore). Fasting glucose was measured in whole blood using a HemoCue 201 RT analyzer (Angelholm, Sweden) upon obtaining capillary blood using a single-use lancet device (Abbott Diabetes Care, Alameda, CA, USA). Volunteers were excluded if they met any one of the following criteria: taken part in sports at the competitive and/or endurance levels; smoking; being allergic to common food (e.g., eggs, shellfish, dairy, nuts, gluten) and test ingredients; intentionally restricting food intake; have glucose-6-phosphate deficiency (G6PD) or metabolic diseases, cardiovascular diseases, or disorders involving the gut, liver, or kidney; taking prescribed medication or dietary supplement that affects glycemia or interferes with other study measurements; consume $\geq$6 alcoholic drinks per week; donated blood within 4 weeks of study participation; had poor veins impeding venous access; had a history of severe vasovagal syncope following blood draws; had or was a carrier of a chronic infection (e.g., hepatitis B or C, human immunodeficiency virus); suffering from active tuberculosis (TB) or receiving treatment for TB; or was related to a member of the study team. All volunteers provided written informed consent. All volunteers were given a choice of attending five or nine test sessions. This study was reviewed and approved by the National Healthcare Group (NHG) Domain-Specific Review Board, Singapore (DRSB Ref: 2018/00258). The study has been registered on clinicaltrials.gov under study ID No. NCT03685916.

2.2. Preparation of Aqueous Extracts and Study Meals

Dried, mature corn silk, commonly sold as an ingredient to treat diabetes by traditional Chinese medicine (TCM) practitioners, was purchased locally. Aqueous extracts were obtained by adding 552 g of corn silk to 2.3 L of tap water and boiling for 15 min. A total of three batches of extraction were undertaken. After pooling, the total volume of the corn silk extract (CS) was made up to 6.9 L and filtered through a fine mesh strainer to remove corn silk particles. Then, 50 mL of this concentrated extract (24 g/100 mL) was aliquoted and stored frozen in a $-20\,^{\circ}$C freezer until further use. An aqueous extract of cumin was prepared by adding 552 g of ground cumin powder (Pattu brand, Sabi Foods (India) Pvt. Ltd., Tamin Nadu, India) into 2.4 L of tap water and allowing the mixture to boil for 15 min. A further 2.4 L of tap water was added to the viscous mixture and boiled for an additional 15 min. Finally, the cumin extract (CU) was filtered through a fine mesh filter and the volume made up to 2.3 L and filtered through a fine mesh strainer to remove cumin particles. Then, 50 mL of this concentrated extract (24 g/100 mL) was aliquoted and stored frozen in a $-20\,^{\circ}$C freezer until further use. For the tamarind extract (TAM), 460 g of deseeded tamarind pulp (Lion Dates Impex Pte Ltd., Trichy, India) was added to 1.2 L of tap water and boiled for 15 min. The first aqueous extract was obtained by filtering the mixture through a fine mesh filter and the residual tamarind pulp re-extracted with another 1.2 L of tap water via boiling for another 15 min. The mixture from the second extraction was filtered and both tamarind aqueous extracts were pooled and the volume made up to 2.3 L. Then, 50 mL of this concentrated extract (20 g/100 mL) was aliquoted and stored frozen in a $-20\,^{\circ}$C freezer until further use. All stored concentrated extracts were thawed and diluted either four-fold for CS-12, CU, and TAM, or diluted two-fold for CS-24 with filtered water on the day of the test session, to make up to a total of 200 mL of "diluted aqueous extract".

The study meals (including rice) were prepared fresh every morning, and the test extracts were either added into rice (R) during cooking or made up into a drink (D). The study meals were rich in carbohydrates, providing approximately 50 g of available carbohydrates mainly from high-GI glutinous rice (Thai glutinous rice, Fairprice, Singapore) and 20 g of cooked garden peas (Frozen garden peaks, Fairprice, Singapore). The control meal (CON) was made up of 65 g of glutinous rice cooked in 200 mL of filtered water in a household rice cooker, 20 g of peas (steamed for 2 min in microwave), and a glass (200 mL) of filtered water. The rice (R) test meals consisted of 60 g of glutinous rice cooked in 200 mL of the diluted aqueous extract (detailed above), 20 g of peas, and 200 mL of filtered water, whereas the drink (D) test meals consisted of 60 g of glutinous rice cooked in 200 mL of plain filtered water, 20 g of peas, and 200 mL of diluted aqueous extract. The food ingredients used and the energy content and nutrient composition of each test meal as obtained from food packet labels or nutrient databases (US Department of Agriculture) are shown in Table 1.

2.3. Total Polyphenol Content (TPC Analyses) of Aqueous Extracts

Total polyphenol contents (TPCs) were determined by the Folin–Ciocalteu (FC) assay in triplicates using a method adopted from Medina-Remón et al. [24]. Briefly, 1 mL of the aqueous concentrated extract was transferred to a microfuge tube and centrifuged at $13,000\times g$ for 5 min at room temperature to obtain the supernatant. Subsequently, 100 µL of tap water (blank), supernatant, or gallic acid standard was added to 200 µL of 10% FC reagent in a 96-well clear polystyrene microplate followed by the addition of 800 µL of 700 mM sodium carbonate. After a 2-h incubation at room temperature in the dark, the absorbance at 765 nm was measured in a microplate reader (Tecan Infinite 200, Mannedorf, Switzerland). The TPC was calculated using the regression equation derived from a gallic acid standard curve and expressed as gallic acid equivalents (GAE).

2.4. Study Design

This was a randomized crossover design whereby each volunteer came for five or nine test sessions with a minimum 1-day wash-out between sessions. The order of the study meals were randomized using the RAND function in Microsoft Excel. Volunteers were asked to avoid strenuous exercise and alcohol intake 24 h prior to their test session and to arrive at the center at 08:30 after an overnight fast of at least 10 h. Upon arrival, volunteers had their baseline (T0) blood pressure measured and a fasted blood sample drawn (after 5 min of rest) at the antecubital vein via cannula by a trained phlebotomist. Volunteers were then served the study meal and instructed to consume the meal within 15 min. Subsequent blood collection was carried out at T15, 30, 45, 60, 90, 120, 150, and 180 min after the first bite. Duplicate blood pressure measurements were taken hourly at T60, 120, and 180 min. A schematic of the study design is shown in Figure 1.

Table 1. Food ingredients, energy content, and nutrient composition of the test meals.

Treatment	Ingredients (per test meal)					Energy or Nutrients (per test meal)					
	Rice (g)	Peas (g)	Dried corn silk (g)	Cumin powder (g)	Tamarind pulp (g)	Energy (kcal)	Carbohydrate (g)	Protein (g)	Fat (g)	Fiber (g)	Available CHO (g)
CON	65	20	-	-	-	249.70	51.56	6.65	1.60	1.77	49.79
CS12	60	20	12	-	-	248.51	51.12	6.95	1.52	2.44	48.68
CS24	60	20	24	-	-	265.01	54.45	7.65	1.57	3.16	51.29
CU	60	20	-	24	-	288.16	53.53	8.29	4.26	4.11	49.42
TAM	60	20	-	-	20	259.00	52.79	6.87	1.98	2.13	50.65

CHO: carbohydrates; CON: control; CS12: corn silk 12 g; CS24: corn silk 24 g; CU: cumin; TAM: tamarind.

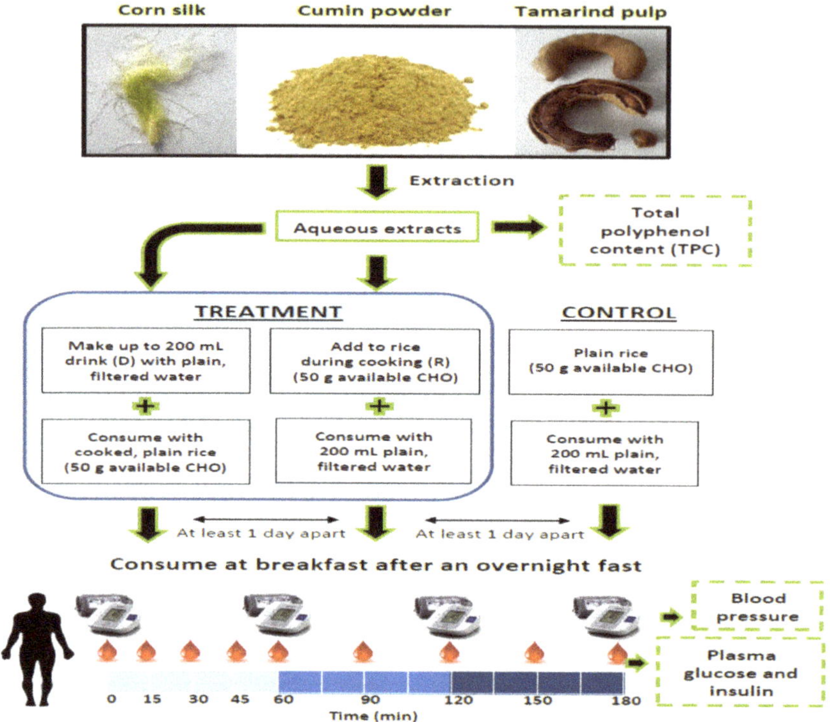

Figure 1. Schematic of study design.

2.5. Blood Sampling and Analytical Methods

Venous blood was collected in K_2EDTA vacutainers (BD, New Jersey, NJ, USA), placed immediately on ice, and centrifuged at $1500 \times g$ for 15 min within an hour of blood draw to obtain plasma. Plasma samples were then stored at −80 °C in aliquots until ready for analyses. Plasma glucose and insulin concentrations were determined by the COBAS c311 and e411 automated analyzers (Roche Diagnostics GmbH, Berlin, Germany), respectively. For each analyzer, a successful two-point calibration was performed at the start of the study. Thereafter, quality control was done daily before analyzing the plasma samples. For each assay, 250 µL of thawed undiluted plasma was analyzed in a Hitachi cup using the manufacturer's recommended cassette for glucose and insulin measurement. All samples were analyzed over a 5-day period using the same analytical instrumentations and laboratory consumables (including standards and quality controls). All samples from a given individual (i.e., all treatments and all time points) were analyzed within the same batch and the same day, in order to limit any effects of experimental variations. The inter-batch coefficient of variation for glucose analysis was 7.51%, and that for insulin analysis was 1.99%.

2.6. Statistical Analyses

The primary outcome of this study was the acute effects of the test ingredients on the postprandial glucose and insulin responses for up to 180 min. The secondary outcomes were: i) the effect of the form in which the test ingredients were administered (R vs. D) on the 30, 120, and 180 min postprandial glucose and insulin responses; and ii) the acute effects of the test ingredients on the postprandial blood pressure measured over 180 min. Incremental areas under the curve (iAUCs) of glucose and insulin were calculated using the change in glucose or insulin above baseline fasting concentration, ignoring the area beneath the baseline. Total areas under the curve (tAUCs) of systolic and diastolic blood pressure measurements were calculated using the absolute blood pressure values measured at the time points.

All analyses in this study were done using Statistical Package for the Social Sciences (SPSS) version 24 (IBM, New York, NY, USA). Data were assessed for normality using the Shapiro–Wilk test as well as visually from the histogram and normal Q–Q (quantile-quantile) plots. Square-root transformation of the data was used where necessary to achieve normality. The treatment effect on the iAUCs of insulin and glucose was tested using linear mixed effects procedure in SPSS with treatment as the fixed factor and compound symmetry (CS) covariance structure. For the interaction between form and treatment test, a linear mixed effects procedure with both treatment and form as fixed factors was used with CS covariance structure.

3. Results

Mean total polyphenol content (TPC) was lowest in the cumin aqueous extract (103.0 ± 3.4 mg GAE/100 g of cumin powder) and highest in the tamarind aqueous extract (440.1 ± 13.3 mg GAE/100 g of tamarind pulp). Mean TPC of corn silk was 233.1 ± 2.2 mg GAE/100 g of dried corn silk.

Of the 18 volunteers enrolled in the study, 16 volunteers completed all nine test sessions. One volunteer only completed three test sessions, while another only completed five sessions. The baseline anthropometric and metabolic characteristics of the volunteers to indicate the profile of volunteers who completed the study are presented in Table 2. All participants had their fasted glucose concentration at screening visit of less than 6.0 mmol/L, except for one participant who had a concentration of 6.20 mmol/L and would therefore be considered as having impaired fasted glucose (IFG).

Table 2. Baseline anthropometric and metabolic characteristics.

Measurement	Mean $\pm$ S.D.
Age (Years)	37.5 ± 12.5
Height (m)	1.7 ± 0.06
Weight (kg)	63.9 ± 7.09
Body mass index (kg/m^2)	21.8 ± 1.67
Body fat (%)	16.5 ± 4.25
Waist (cm)	78.8 ± 5.36
Systolic blood pressure (mmHg)	120.0 ± 7.75
Diastolic blood pressure (mmHg)	76.5 ± 7.37
Resting heart rate (bpm)	$64. \pm 10.03$
Fasting blood glucose (mmol/L)	5.2 ± 0.43

3.1. Postprandial Glycemic and Insulinemic Response to Control and Test Meals

Figures 2 and 3 respectively show plasma glucose and insulin concentrations over time. As expected, mean plasma glucose concentration rose significantly above baseline fasting levels in the first hour after meal initiation, peaking at T30 and falling below baseline at T150 and T180 (Figure 2). Similarly, the mean plasma insulin concentration rose significantly above baseline fasting levels in the first 2 h after meal initiation, peaking at T45 and returning to baseline levels at T150 (Figure 3). There were no significant differences between the various treatments as compared with the control on postprandial glycemia, irrespective of the food form, evaluated using incremental areas under the curve for various time intervals (iAUC$_{30}$, iAUC$_{120}$, or iAUC$_{180}$), as shown in Table 3. There were also no differences in the postprandial glycemic responses between the two different forms (rice vs. drink) of consumption, irrespective of the test ingredients. Similarly, there were no significant differences in the influence of the various treatments as compared with the control on postprandial insulinemia as evaluated using incremental areas under the curve for various time intervals (iAUC$_{30}$, iAUC$_{120}$, or iAUC$_{180}$). However, there was a significant increase ($p < 0.01$) in postprandial insulinemia (iAUC$_{120}$ or iAUC$_{180}$) when the test ingredients were consumed as a drink (D) as compared with the same ingredients being cooked within rice (R), irrespective of the test ingredients. It should also be noted that the form × treatment interactions were not significant across all measures, as shown in Table 3.

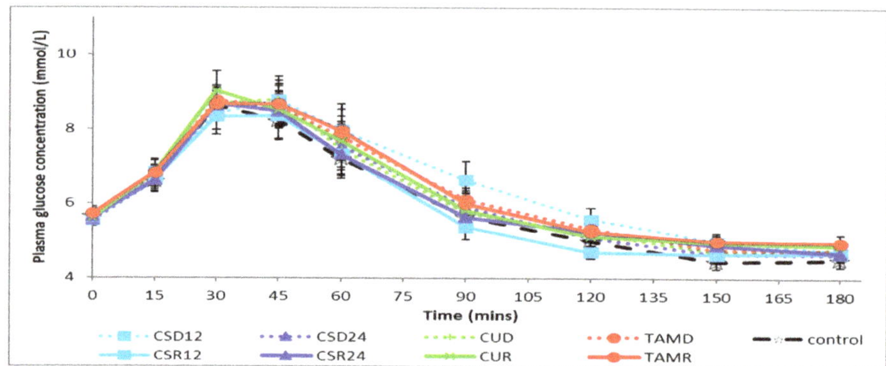

Figure 2. Postprandial changes in plasma glucose concentration during various treatments. CSD12: corn silk drink 12 g; CSR12: corn silk rice 12 g; CSD24: corn silk drink 24 g; CSR24: corn silk rice 24 g; CUD: cumin drink; CUR: cumin rice; TAMD: tamarind drink; TAMR: tamarind rice.

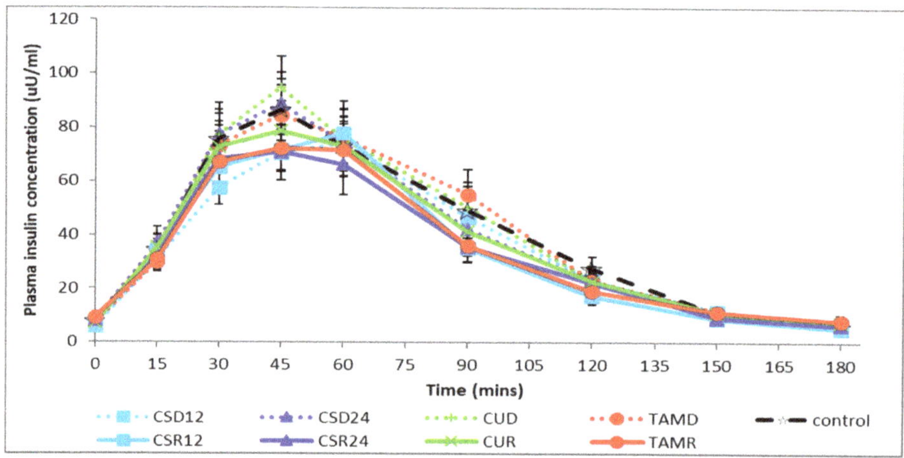

Figure 3. Postprandial changes in plasma insulin concentration during various treatments. CSD12: corn silk drink 12 g; CSR12: corn silk rice 12 g; CSD24: corn silk drink 24 g; CSR24: corn silk rice 24 g; CUD: cumin drink; CUR: cumin rice; TAMD: tamarind drink; TAMR: tamarind rice.

Table 3. Incremental areas under the curve (iAUCs) for postprandial glucose and insulin over relevant time periods during various treatments.

Test Meal	iAUC 0–30 min		iAUC 0–120 min		iAUC 0–180 min	
	Glucose (mmol min/L)	Insulin (µU min/mL)	Glucose (mmol min/L)	Insulin (µU min/mL)	Glucose (mmol min/L)	Insulin (µU min/mL)
Control (plain rice with plain water)	36.55 ± 5.08	908.43 ± 141.98	158.06 ± 29.17	5598.56 ± 708.59	159.91 ± 29.59	6008.20 ± 751.33
Test ingredients cooked in rice and consumed with plain water (R)						
Corn silk (12 g)	36.15 ± 6.77	836.69 ± 128.20	147.99 ± 28.12	4802.84 ± 583.04	148.16 ± 28.11	5028.97 ± 633.59
Corn silk (24 g)	37.71 ± 4.18	839.87 ± 113.04	165.37 ± 30.87	4570.14 ± 543.70	171.41 ± 33.18	4890.21 ± 598.19
Cumin	44.12 ± 7.04	915.75 ± 106.42	193.05 ± 30.72	5223.44 ± 720.45	205.80 ± 33.69	5627.36 ± 817.57
Tamarind	38.97 ± 5.54	764.41 ± 110.33	181.02 ± 28.10	4503.82 ± 558.78	183.04 ± 28.37	4773.23 ± 591.69
Test ingredients added into water and consumed with plain rice (D)						
Corn silk (12 g)	40.98 ± 6.37	768.77 ± 101.19	220.90 ± 41.98	5138.30 ± 681.74	230.45 ± 45.09	5591.53 ± 723.78
Corn silk (24 g)	39.30 ± 5.37	944 ± 148.22	183.03 ± 29.20	5386.49 ± 698.89	186.99 ± 30.32	5729.12 ± 724.89
Cumin	38.54 ± 5.32	893.25 ± 96.40	167.52 ± 29.34	5768.09 ± 676.85	169.21 ± 29.96	6085.28 ± 714.54
Tamarind	39.32 ± 5.39	829.59 ± 98.12	182.05 ± 28.01	5679.60 ± 773.40	186.40 ± 29.89	6047.60 ± 814.98
p-Value for treatment (overall) [a,b]	0.912	0.499	0.404	0.080	0.310	0.051
p-Value for interaction [a,c]	0.632	0.601	0.143	0.669	0.091	0.721
p-Value for main effect of form [a,c]	0.973	0.647	0.274	0.005	0.329	0.003
p-Value for main effect of treatment [a,c]	0.932	0.176	0.983	0.247	0.985	0.227

[a] All analyses were done using the linear mixed effects procedure with compound symmetry covariance structure in SPSS. [b] Model testing effects of treatment overall, including control. [c] Model testing effects of form × treatment without control. Bold lettering indicates $p < 0.05$.

3.2. Postprandial Changes in Blood Pressure during Control and Test Meals

The postprandial changes in systolic and diastolic blood pressures are shown in Figures 4 and 5, respectively. There were no significant differences between treatments, irrespective of form or between forms, irrespective of treatments at baseline and in the total area under the curve (tAUC) of the postprandial systolic or diastolic blood pressure, as shown in Table 4.

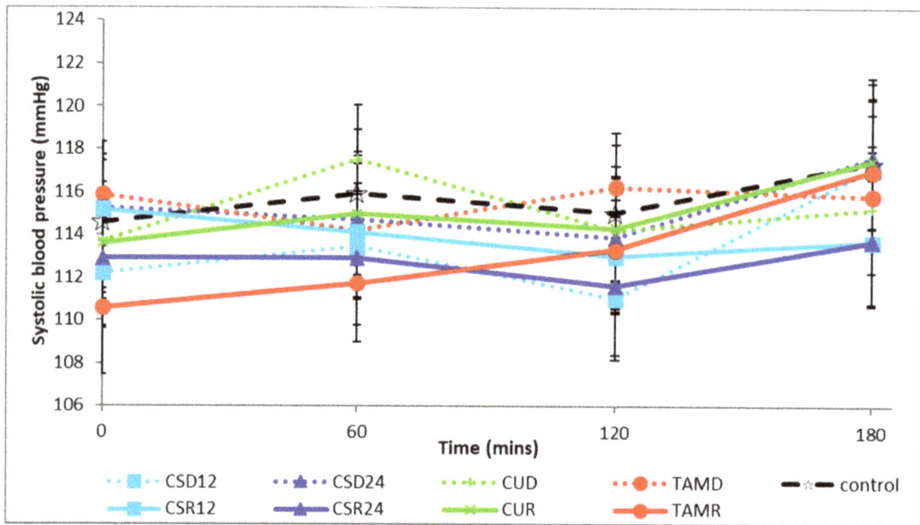

Figure 4. Postprandial changes in systolic blood pressure during various treatments. corn silk drink 12 g; CSR12: corn silk rice 12 g; CSD24: corn silk drink 24 g; CSR24: corn silk rice 24 g; CUD: cumin drink; CUR: cumin rice; TAMD: tamarind drink; TAMR: tamarind rice.

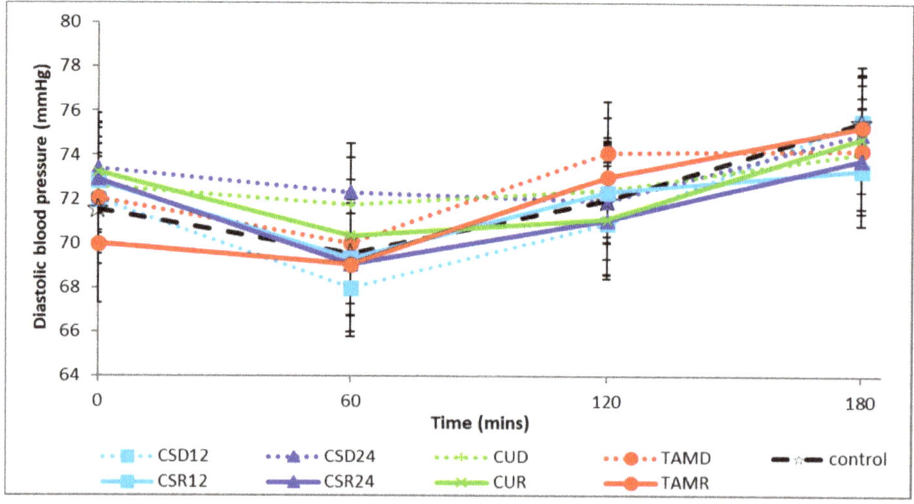

Figure 5. Postprandial changes in diastolic blood pressure during various treatments. CSD12: corn silk drink 12 g; CSR12: corn silk rice 12 g; CSD24: corn silk drink 24 g; CSR24: corn silk rice 24 g; CUD: cumin drink; CUR: cumin rice; TAMD: tamarind drink; TAMR: tamarind rice.

Table 4. Baseline and total areas under the curve (tAUCs) of blood pressure over the postprandial period.

Treatment	Baseline (0 min)		Total AUC 0–180 min	
	Systolic BP (mmHg)	Diastolic BP (mmHg)	Systolic BP (mmHg min)	Diastolic BP (mmHg min)
Control	114.58 ± 3.12	71.56 ± 2.52	20,815.00 ± 528.18	12,902.50 ± 391.60
Test ingredients cooked in rice and consumed with plain water (R)				
Corn silk (12 g)	115.19 ± 2.25	72.81 ± 2.59	20,493.75 ± 458.67	11,454.17 ± 1049.87
Corn silk (24 g)	112.91 ± 3.16	72.91 ± 2.32	20,270.63 ± 526.11	11,385.83 ± 1034.56
Cumin	113.61 ± 2.37	73.22 ± 2.65	20,689.17 ± 471.23	12,929.17 ± 419.73
Tamarind	110.59 ± 3.11	70.0 ± 2.68	20,331.18 ± 487.74	12,169.17 ± 842.99
Test ingredients added into water and consumed with plain rice (D)				
Corn silk (12 g)	112.22 ± 2.57	72.0 ± 2.48	20,349.38 ± 448.14	11,344.17 ± 1030.44
Corn silk (24 g)	115.25 ± 2.46	73.41 ± 2.07	20,701.88 ± 511.97	11,649.17 ± 1046.57
Cumin	113.69 ± 2.74	72.61 ± 2.17	20,769.17 ± 448.02	13,054.17 ± 381.35
Tamarind	115.88 ± 2.44	72.06 ± 1.84	20,777.65 ± 365.70	12,315.83 ± 801.08
p-Value for treatment (overall)	0.206	0.705	0.624	0.393

4. Discussion

Given that Asians are at a greater risk of type 2 diabetes and prediabetes, and considering that the dietary carbohydrate loads in Asians are much higher than in other parts of the world, there is a drive to identify functional ingredients within the Asian dietary context which can improve glycemic control. This trial was set out to test the effects of three separate ingredients, at dietary doses, for their individual effects on postprandial glycemia and insulinemia when consumed along with a carbohydrate-rich meal consisting of a portion of high-GI rice. We have previously shown improvements in postprandial glycemic control with polyphenol-rich mixed spices [25,26]. The three ingredients chosen in this study were also rich in phenolic compounds as measured here and reported elsewhere for cumin [27–29], tamarind [30–33], and dried corn silk [5,34]. Furthermore, inhibition of α-amylase and α-glucosidase activities are important determinants for any natural compound with a potential to reduce postprandial glycemia [35], and all three ingredients used in our study (or compounds found within them) have been shown to possess such activities in previous studies [8,36–39]. These ingredients were also specifically chosen for their extensive traditional utility, particularly across Asia. Cumin and tamarind are both used in several preparations as spices/condiments in rice-based dishes (e.g., *"jeera"* (cumin) rice), curries, soups (e.g., cumin and/or tamarind in *"rasam"*, mulligatawny soup, tom yum goong soups, etc.) or as beverages such as *"jal jeera"* (cumin water in India) or as *"nam makham"* (tamarind drink in Thailand). Given their extensive consumption in several different forms (i.e., in foods and in beverages), we further investigated any potential differences in their glycemic/insulinemic responses when either consumed as a drink or added into rice while cooking. Corn silk extract (CS) on the other hand was investigated not only due to its extensive use by traditional Chinese medicine practitioners as an anti-diabetic formulation [4,40], but also because corn silk is a common agricultural waste product which is rich in polyphenols and there is potential utility of its reintroduction back into the food system. This is particularly relevant considering the recent drive to re-introduce polyphenol-rich food-processing by-products back into the food system, as had been discussed in detail elsewhere [41].

Despite our careful selection of the ingredients tested, our study failed to demonstrate any significant effects on postprandial glycemia or insulinemia as compared to the control meal without these ingredients. There could be several reasons for this observation. First of all, we recruited non-diabetic, healthy individuals, all of whom are likely to have had adequate glycemic control. Indeed, some studies have previously shown that the postprandial glycemic response even between high- and low-GI meals did not differ in young healthy individuals [42]. Furthermore, the postprandial metabolic responses to the same foods are often different between normal and at-risk individuals [43,44]. Therefore, the results of our study may not be applicable to type 2 diabetic or other at-risk populations, who may well have benefited from the ingredients used. Nonetheless, the strengths of our study were that we used a crossover design, carefully matched the macronutrient composition between treatments, and used dietary doses of the ingredients in a real-life dietary context. In contrast, most of the animal studies supporting favorable effects often used much higher relative doses which are not achievable through usual dietary means in humans. Nonetheless, considering that our ingredients were rich in various phytochemicals with antioxidant as well as anti-inflammatory properties, there may be other favorable downstream effects including the prevention of complications associated with postprandial hyperglycemia (especially in type 2 diabetics), including oxidative stress, chronic inflammation, etc., which were not measured in this study.

While the phytochemical-rich ingredients used in our study per se did not give rise to any obvious effects on postprandial glycemia or insulinemia, there was a significant lowering of the postprandial insulinemic responses when the test meals were consumed within cooked rice (R) as compared to being consumed in the drink (D) form separately from the rice. This suggests that the food matrix in which phytochemical-rich ingredients are consumed has an effect on postprandial metabolic responses. Our research group has previously shown that both postprandial glycemic [22] and lipidemic [45] responses depended on the food matrix. Others have also shown that the postprandial glycemic lowering potential of polyphenol-rich ingredients depended on the form in which it was consumed [46], and various cooking methods were shown to significantly alter polyphenol bioaccessibility as well as the biological properties of polyphenol-rich ingredients, including differences in α-glucosidase inhibitory activity [47]. Furthermore, we and others have reported that the timing/sequence of consumption of polyphenol-rich ingredients (e.g., consumption prior to rather than concurrent consumption) produced different effects [48,49].

Similar to our findings, others have also reported greater postprandial insulinemic responses to a homogenized meal as compared with a solid meal [23]. There could be several mechanistic explanations for such observations. It is well known that various polyphenols can interact with starch molecules, leading to the formation of starch–polyphenol complexes [50,51] which can significantly restrict the access of digestive enzymes (e.g., amylases and glucosidases) and thereby the in vitro and in vivo digestibility of these starches [52–54]. Thus, the prior cooking of our ingredients within rice (R) may have led to differences in the extent of starch–polyphenol complex formations as well as starch accessibility and digestibility as compared to the ingredients being co-consumed within a separate drink (D). Moreover, we and others have previously shown that polyphenol-rich foods, including spices, can increase the secretion of gut hormones such as glucagon-like peptide-1 (GLP-1) [26,55]. It has also been shown previously that liquid meals (prepared by homogenizing solid meals) produced greater levels of both postprandial GLP-1 and insulin as compared to the same meal consumed solid [56]. Given that the gastric emptying of the liquid part of mixed (solid–liquid) meals occurs before the solid part [57], this may further explain the relatively greater insulinemic response when the ingredients were consumed within a drink (D) as compared to the same consumed within cooked rice (R). Given that postprandial insulinemia is an independent risk factor for diabetes and cardiovascular diseases [58,59], findings that food form plays a role has potential clinical implications. More research is therefore needed to confirm this observation.

5. Conclusions

This study demonstrated that adding aqueous extracts of corn silk, cumin, and tamarind at dietary doses, in two separate forms, to a high-GI rice consisting of approximately 50 g of available carbohydrates in healthy volunteers conferred no additional benefits on postprandial glycemia or insulinemia. Our randomized, controlled clinical trial findings are insightful and advantageous in that these ingredients are traditionally believed in Asia to be beneficial in reducing the risk of diabetes, although to date, the majority of the previous evidence supporting these claims had been undertaken in animals, most often at relatively high doses. Some of the limitations of this study were the acute nature of the study design, the limited number of postprandial parameters being measured, and the fact that this study was undertaken in a cohort of healthy, normoglycemic, non-diabetic individuals. While other longer-term beneficial effects (e.g., anti-inflammatory, antioxidant, etc.) of the test ingredients in a population with a compromised glucose homeostasis (e.g., type 2 diabetics) cannot be ruled out, this study highlights the need for more controlled clinical trials in the future before specific claims are assumed regarding the specific benefits of traditional ingredients. Finally, this study also highlights the importance of the food form on postprandial insulinemia, in that phytochemical-rich ingredients in liquid forms (e.g., soups and beverages) may elicit greater postprandial insulinemic response than when consumed within a solid meal. Therefore, future applications of our study findings include the manipulation of food forms via various food consumption, preparation, and processing methods, in order to improve the metabolic consequences of certain foods and/or composited dishes.

Author Contributions: Conceptualization, S.H., L.G., and C.J.H.; methodology, S.H., L.G., S.P., and S.L.T.; statistical software, S.P.; validation, S.H., L.G., and S.P.; formal analysis, S.H., L.G., and S.P.; investigation, S.H., L.G., and S.L.T.; resources, S.H., L.G., and S.L.T.; data curation, S.H., L.G., and S.P.; writing—original draft preparation, S.H. and L.G.; writing—review and editing, S.H, L.G., S.L.T., and C.J.H.; visualization, S.P. and S.L.T.; supervision, S.H.; project administration, S.H., C.J.H.; funding acquisition, C.J.H.

Funding: This research was funded by the Singapore Institute for Clinical Sciences, Singapore, and the APC was funded by the Singapore Institute for Clinical Sciences, Singapore.

Acknowledgments: We thank Joseph Lim, Sze Han Lee, and Susanna Lim Poh Suan for their technical assistance. We thank all volunteers for taking part in this study.

Conflicts of Interest: The authors declare no conflicts of interest.

References

1. Krishnaswamy, K. Traditional Indian spices and their health significance. *Asia Pac. J. Clin. Nutr.* **2008**, *17*, 265–268.

2. Rizvi, S.I.; Mishra, N. Traditional Indian medicines used for the management of diabetes mellitus. *J. Diabetes Res.* **2013**, *2013*, 11. [CrossRef] [PubMed]

3. Li, W.; Zheng, H.; Bukuru, J.; De Kimpe, N. Natural medicines used in the traditional Chinese medical system for therapy of diabetes mellitus. *J. Ethnopharmacol.* **2004**, *92*, 1–21. [CrossRef] [PubMed]

4. Wang, B.; Xiao, T.; Ruan, J.; Liu, W. Beneficial effects of corn silk on metabolic syndrome. *Curr. Pharm. Des.* **2017**, *23*, 5097–5103. [CrossRef] [PubMed]

5. Hasanudin, K.; Hashim, P.; Mustafa, S. Corn silk (Stigma maydis) in healthcare: A phytochemical and pharmacological review. *Molecules* **2012**, *17*, 9697–9715. [CrossRef]

6. Kai-Jin Wang, J.-L.Z. Corn silk (*Zea mays* L.), a source of natural antioxidants with α-amylase, α-glucosidase, advanced glycation and diabetic nephropathy inhibitory activities. *Biomed. Pharmacother.* **2019**, *110*, 510–517. [CrossRef] [PubMed]

7. Guo, Q.; Chen, Z.; Santhanam, R.K.; Xu, L.; Gao, X.; Ma, Q.; Xue, Z.; Chen, H. Hypoglycemic effects of polysaccharides from corn silk (*Maydis stigma*) and their beneficial roles via regulating the PI3K/Akt signaling pathway in L6 skeletal muscle myotubes. *Int. J. Biol. Macromol.* **2019**, *121*, 981–988. [CrossRef]

8. Sabiu, S.; O'neill, F.; Ashafa, A. Kinetics of α-amylase and α-glucosidase inhibitory potential of *Zea mays* Linnaeus (Poaceae), Stigma maydis aqueous extract: An in vitro assessment. *J. Ethnopharmacol.* **2016**, *183*, 1–8. [CrossRef]

9. Ho, T.-Y.; Li, C.-C.; Lo, H.-Y.; Chen, F.-Y.; Hsiang, C.-Y. Corn silk extract and its bioactive peptide ameliorated lipopolysaccharide-induced inflammation in mice via the nuclear factor-κB signaling pathway. *J. Agric. Food Chem.* **2017**, *65*, 759–768. [CrossRef]

10. Chang, C.-C.; Yuan, W.; Roan, H.-Y.; Chang, J.-L.; Huang, H.-C.; Lee, Y.-C.; Tsay, H.J.; Liu, H.-K. The ethyl acetate fraction of corn silk exhibits dual antioxidant and anti-glycation activities and protects insulin-secreting cells from glucotoxicity. *BMC Complement. Altern. Med.* **2016**, *16*, 432. [CrossRef]

11. Liu, J.; Wang, C.; Wang, Z.; Zhang, C.; Lu, S.; Liu, J. The antioxidant and free-radical scavenging activities of extract and fractions from corn silk (*Zea mays* L.) and related flavone glycosides. *Food Chem.* **2011**, *126*, 261–269. [CrossRef]

12. Willatgamuwa, S.; Platel, K.; Saraswathi, G.; Srinivasan, K. Antidiabetic influence of dietary cumin seeds (*Cuminum cyminum*) in streptozotocin induced diabetic rats. *Nutr. Res.* **1998**, *18*, 131–142. [CrossRef]

13. Dhandapani, S.; Subramanian, V.R.; Rajagopal, S.; Namasivayam, N. Hypolipidemic effect of *Cuminum cyminum* L. on alloxan-induced diabetic rats. *Pharmacol. Res.* **2002**, *46*, 251–255. [CrossRef]

14. Roman-Ramos, R.; Flores-Saenz, J.; Alarcon-Aguilar, F. Anti-hyperglycemic effect of some edible plants. *J. Ethnopharmacol.* **1995**, *48*, 25–32. [CrossRef]

15. Lee, H.-S. Cuminaldehyde: Aldose reductase and α-glucosidase inhibitor derived from *Cuminum cyminum* L. seeds. *J. Agric. Food Chem.* **2005**, *53*, 2446–2450. [CrossRef] [PubMed]

16. Azman, K.F.; Amom, Z.; Azlan, A.; Esa, N.M.; Ali, R.M.; Shah, Z.M.; Kadir, K.K.A. Antiobesity effect of *Tamarindus indica* L. pulp aqueous extract in high-fat diet-induced obese rats. *J. Nat. Med.* **2012**, *66*, 333–342. [CrossRef] [PubMed]

17. Maiti, R.; Das, U.K.; Ghosh, D. Attenuation of hyperglycemia and hyperlipidemia in streptozotocin-induced diabetic rats by aqueous extract of seed of Tamarindus indica. *Biol. Pharm. Bulletin* **2005**, *28*, 1172–1176. [CrossRef] [PubMed]

18. Maiti, R.; De, D.; Ghosh, D. Antidiabetic effect of n-hexane fraction of hydro-methanolic extract of tamarindus indica linn seed in streptozotocin-induced diabetic rat: A correlative approach with in vivo and in vitro antioxidant activities. *Int. J. Pharm. Sci. Res.* **2018**, *9*, 1821–1830.

19. Maiti, R.; Jana, D.; Das, U.; Ghosh, D. Antidiabetic effect of aqueous extract of seed of Tamarindus indica in streptozotocin-induced diabetic rats. *J. Ethnopharmacol.* **2004**, *92*, 85–91. [CrossRef]

20. Bhadoriya, S.S.; Ganeshpurkar, A.; Bhadoriya, R.P.S.; Sahu, S.K.; Patel, J.R. Antidiabetic potential of polyphenolic-rich fraction of Tamarindus indica seed coat in alloxan-induced diabetic rats. *J. Basic Clin. Physiol. Pharmacol.* **2018**, *29*, 37–45. [CrossRef]

21. Zhu, R.; Liu, M.; Han, Y.; Wang, L.; Ye, T.; Lu, J.; Fan, Z. Acute effects of non-homogenised and homogenised vegetables added to rice-based meals on postprandial glycaemic responses and in vitro carbohydrate digestion. *Br. J. Nutr.* **2018**, *120*, 1023–1033. [CrossRef]

22. Tey, S.; Lee, D.; Henry, C.J. Fruit form influences postprandial glycemic response in elderly and young adults. *J. Nutr. Health Aging* **2017**, *21*, 887–891. [CrossRef] [PubMed]

23. Peracchi, M.; Santangelo, A.; Conte, D.; Fraquelli, M.; Tagliabue, R.; Gebbia, C.; Porrini, M. The physical state of a meal affects hormone release and postprandial thermogenesis. *Br. J. Nutr.* **2000**, *83*, 623–628. [CrossRef]

24. Medina-Remón, A.; Barrionuevo-González, A.; Zamora-Ros, R.; Andres-Lacueva, C.; Estruch, R.; Martínez-González, M.-Á.; Diez-Espino, J.; Lamuela-Raventos, R.M. Rapid Folin-Ciocalteu method using microtiter 96-well plate cartridges for solid phase extraction to assess urinary total phenolic compounds, as a biomarker of total polyphenols intake. *Analytica Chimica Acta* **2009**, *634*, 54–60. [CrossRef] [PubMed]

25. Haldar, S.; Chia, S.C.; Lee, S.H.; Lim, J.; Leow, M.K.-S.; Chan, E.C.Y.; Henry, C.J. Polyphenol-rich curry made with mixed spices and vegetables benefits glucose homeostasis in Chinese males (polyspice study): A dose-response randomized controlled crossover trial. *Eur. J. Nutr.* **2019**, *58*, 301–313. [CrossRef] [PubMed]

26. Haldar, S.; Chia, S.C.; Henry, C.J. Polyphenol-rich curry made with mixed spices and vegetables increases postprandial plasma GLP-1 concentration in a dose-dependent manner. *Eur. J. Clin. Nutr.* **2018**, *72*, 297. [CrossRef] [PubMed]

27. Srinivasan, K. Cumin (*Cuminum cyminum*) and black cumin (*Nigella sativa*) seeds: Traditional uses, chemical constituents, and nutraceutical effects. *Food Qual. Saf.* **2018**, *2*, 1–16. [CrossRef]

28. Zhang, Y.; Ma, H.; Liu, W.; Yuan, T.; Seeram, N.P. New antiglycative compounds from cumin (*Cuminum cyminum*) spice. *J. Agric. Food Chem.* **2015**, *63*, 10097–10102. [CrossRef] [PubMed]

29. Mnif, S.; Aifa, S. Cumin (*Cuminum cyminum* L.) from traditional uses to potential biomedical applications. *Chem. Biodivers.* **2015**, *12*, 733–742. [CrossRef]

30. Ishaku, G.A.; Ardo, B.P.; Abubakar, H.; Peingurta, F.A. Nutritional composition of tamarindus indica fruit pulp. *J. Chem. Sci.* **2016**, *6*, 695–699.

31. Razali, N.; Mat-Junit, S.; Abdul-Muthalib, A.F.; Subramaniam, S.; Abdul-Aziz, A. Effects of various solvents on the extraction of antioxidant phenolics from the leaves, seeds, veins and skins of *Tamarindus indica* L. *Food Chem.* **2012**, *131*, 441–448. [CrossRef]

32. Bhadoriya, S.S.; Ganeshpurkar, A.; Narwaria, J.; Rai, G.; Jain, A.P. *Tamarindus indica*: Extent of explored potential. *Pharmacognosy Rev.* **2011**, *5*, 73.

33. Da Silva, L.M.R.; de Figueiredo, E.A.T.; Ricardo, N.M.P.S.; Vieira, I.G.P.; de Figueiredo, R.W.; Brasil, I.M.; Gomes, C.L. Quantification of bioactive compounds in pulps and by-products of tropical fruits from Brazil. *Food Chem.* **2014**, *143*, 398–404. [CrossRef] [PubMed]

34. Tian, J.; Chen, H.; Chen, S.; Xing, L.; Wang, Y.; Wang, J. Comparative studies on the constituents, antioxidant and anticancer activities of extracts from different varieties of corn silk. *Food Funct.* **2013**, *4*, 1526–1534. [CrossRef] [PubMed]

35. Tarling, C.A.; Woods, K.; Zhang, R.; Brastianos, H.C.; Brayer, G.D.; Andersen, R.J.; Withers, S.G. The search for novel human pancreatic α-amylase inhibitors: High-throughput screening of terrestrial and marine natural product extracts. *ChemBioChem* **2008**, *9*, 433–438. [CrossRef]

36. Siow, H.-L.; Tye, G.-J.; Gan, C.-Y. Pre-clinical evidence for the efficacy and safety of α-amylase inhibitory peptides from cumin (*Cuminum cyminum*) seed. *J. Funct. Foods* **2017**, *35*, 216–223. [CrossRef]

37. Chen, S.; Chen, H.; Tian, J.; Wang, Y.; Xing, L.; Wang, J. Chemical modification, antioxidant and α-amylase inhibitory activities of corn silk polysaccharides. *Carbohydr. Polym.* **2013**, *98*, 428–437. [CrossRef]

38. Adisakwattana, S.; Chantarasinlapin, P.; Thammarat, H.; Yibchok-Anun, S. A series of cinnamic acid derivatives and their inhibitory activity on intestinal α-glucosidase. *J. Enzyme Inhib. Med. Chem.* **2009**, *24*, 1194–1200. [CrossRef]

39. Pereira, D.F.; Cazarolli, L.H.; Lavado, C.; Mengatto, V.; Figueiredo, M.S.R.B.; Guedes, A.; Pizzolatti, M.G.; Silva, F.R.M.B. Effects of flavonoids on α-glucosidase activity: Potential targets for glucose homeostasis. *Nutrition* **2011**, *27*, 1161–1167. [CrossRef]

40. Wang, N.; Zhu, F.; Chen, K. Advances in Chinese herbal medicine for the treatment of diabetes. *Int J. Clin. Exp. Med.* **2017**, *10*, 13025–13036.

41. De Camargo, A.; Schwember, A.; Parada, R.; Garcia, S.; Maróstica Júnior, M.; Franchin, M.; Regitano-d'Arce, M.; Shahidi, F. Opinion on the hurdles and potential health benefits in value-added use of plant food processing by-products as sources of phenolic compounds. *Int. J. Mol. Sci.* **2018**, *19*, 3498. [CrossRef] [PubMed]

42. Kahlhöfer, J.; Karschin, J.; Silberhorn-Bühler, H.; Breusing, N.; Bosy-Westphal, A. Effect of low-glycemic-sugar-sweetened beverages on glucose metabolism and macronutrient oxidation in healthy men. *Int. J. Obes.* **2016**, *40*, 990. [CrossRef] [PubMed]

43. Anderson, R.A.; Evans, M.L.; Ellis, G.R.; Graham, J.; Morris, K.; Jackson, S.K.; Lewis, M.J.; Rees, A.; Frenneaux, M.P. The relationships between post-prandial lipaemia, endothelial function and oxidative stress in healthy individuals and patients with type 2 diabetes. *Atherosclerosis* **2001**, *154*, 475–483. [CrossRef]

44. Wolever, T.M.; Campbell, J.E.; Geleva, D.; Anderson, G.H. High-fiber cereal reduces postprandial insulin responses in hyperinsulinemic but not normoinsulinemic subjects. *Diabetes Care* **2004**, *27*, 1281–1285. [CrossRef] [PubMed]

45. Tan, S.-Y.; Peh, E.; Siow, P.C.; Henry, C.J. The Role of Ethylcellulose Oleogel in Human Health and Its Potential Applications. In *Edible Oleogels*; Marangoni, A., Garti, N., Eds.; Elsevier: Amsterdam, The Netherlands, 2018; pp. 401–414.

46. Kerimi, A.; Nyambe-Silavwe, H.; Gauer, J.S.; Tomás-Barberán, F.A.; Williamson, G. Pomegranate juice, but not an extract, confers a lower glycemic response on a high–glycemic index food: Randomized, crossover, controlled trials in healthy subjects. *Am. J. Clin. Nutr.* **2017**, *106*, 1384–1393. [CrossRef]

47. Kumar, D.A.; Anusha, S.V.; Oruganti, S.; Deshpande, M.; Zehra, A.; Tiwari, A.K. Raw versus cooked vegetable juice. *Nutrafoods* **2015**, *14*, 27–38. [CrossRef]

48. Kasuya, N.; Okuyama, M.; Yoshida, K.; Sunabori, S.; Suganuma, H.; Murata, I.; Inoue, Y.; Kanamoto, I. Prior or concomitant drinking of vegetable juice with a meal attenuates postprandial blood glucose elevation in healthy young adults. *Food Nutr. Sci.* **2016**, *7*, 797. [CrossRef]

49. Sun, L.; Goh, H.J.; Govindharajulu, P.; Leow, M.K.-S.; Henry, C.J. Postprandial glucose, insulin and incretin responses differ by test meal macronutrient ingestion sequence (PATTERN study). *Clin. Nutr.* **2019**, *13*, 126–133. [CrossRef]

50. Guo, Z.; Zhao, B.; Chen, J.; Chen, L.; Zheng, B. Insight into the characterization and digestion of lotus seed starch-tea polyphenol complexes prepared under high hydrostatic pressure. *Food Chem.* **2019**, *297*, 124992. [CrossRef]

51. Amoako, D.B.; Awika, J.M. Resistant starch formation through intrahelical V-complexes between polymeric proanthocyanidins and amylose. *Food Chem.* **2019**, *285*, 326–333. [CrossRef]

52. Miao, M.; Jiang, H.; Jiang, B.; Zhang, T.; Cui, S.W.; Jin, Z. Phytonutrients for controlling starch digestion. Evaluation of grape skin extract. *Food Chem.* **2014**, *145*, 205–211. [CrossRef] [PubMed]

53. Barros, F.; Awika, J.M.; Rooney, L.W. Interaction of tannins and other sorghum phenolic compounds with starch and effects on in vitro starch digestibility. *J. Agric. Food Chem.* **2012**, *60*, 11609–11617. [CrossRef] [PubMed]

54. Chai, Y.; Wang, M.; Zhang, G. Interaction between amylose and tea polyphenols modulates the postprandial glycemic response to high-amylose maize starch. *J. Agric. Food Chem.* **2013**, *61*, 8608–8615. [CrossRef] [PubMed]

55. Domínguez Avila, J.; Rodrigo Garcia, J.; González Aguilar, G.; de la Rosa, L. The antidiabetic mechanisms of polyphenols related to increased glucagon-like peptide-1 (GLP1) and insulin signaling. *Molecules* **2017**, *22*, 903. [CrossRef] [PubMed]

56. Brynes, A.E.; Frost, G.S.; Edwards, C.M.B.; Ghatei, M.A.; Bloom, S.R. Plasma glucagon-like peptide-1 (7-36) amide (GLP-1) response to liquid phase, solid phase, and meals of differing lipid composition. *Nutrition* **1998**, *14*, 433–436. [CrossRef]

57. Collins, P.; Houghton, L.; Read, N.; Horowitz, M.; Chatterton, B.; Heddle, R.; Dent, J. Role of the proximal and distal stomach in mixed solid and liquid meal emptying. *Gut* **1991**, *32*, 615–619. [CrossRef] [PubMed]

58. Del Prato, S.; Leonetti, F.; Simonson, D.; Sheehan, P.; Matsuda, M.; DeFronzo, R.A. Effect of sustained physiologic hyperinsulinaemia and hyperglycaemia on insulin secretion and insulin sensitivity in man. *Diabetologia* **1994**, *37*, 1025–1035. [CrossRef] [PubMed]

59. Hsueh, W.A.; Law, R.E. Cardiovascular risk continuum: Implications of insulin resistance and diabetes. *Am. J. Med.* **1998**, *105*, 4S–14S. [CrossRef]

Article

Polyphenolic Characterization, Antioxidant, and Cytotoxic Activities of *Mangifera indica* Cultivars from Costa Rica

Mirtha Navarro [1,*], Elizabeth Arnaez [2], Ileana Moreira [3], Silvia Quesada [3], Gabriela Azofeifa [3], Krissia Wilhelm [1], Felipe Vargas [1] and Pei Chen [4]

[1] Department of Chemistry, University of Costa Rica (UCR), Rodrigo Facio Campus, San Pedro Montes Oca, San Jose 2060, Costa Rica
[2] Department of Biology, Technological University of Costa Rica (TEC), Cartago 7050, Costa Rica
[3] Department of Biochemistry, School of Medicine, University of Costa Rica (UCR), Rodrigo Facio Campus, San Pedro Montes Oca, San Jose 2060, Costa Rica
[4] Methods and Application of Food Composition Laboratory, Beltsville Human Nutrition Research Center, Agricultural Research Service, U.S. Department of Agriculture, Beltsville, MD 20705, USA
* Correspondence: mnavarro@codeti.org; Tel.: +506-8873-5539

Received: 30 June 2019; Accepted: 20 August 2019; Published: 2 September 2019

Abstract: The phenolic profile of skin and flesh from *Manifera indica* main commercial cultivars (Keitt and Tommy Atkins) in Costa Rica was studied using ultra performance liquid chromatography coupled with high resolution mass spectrometry (UPLC-ESI-MS) on enriched phenolic extracts. A total of 71 different compounds were identified, including 32 gallates and gallotannins (of different polymerization degree, from galloyl hexose monomer up to decagalloyl hexoses and undecagalloyl hexoses); seven hydroxybenzophenone (maclurin and iriflophenone) derivatives, six xanthonoids (including isomangiferin and mangiferin derivatives); 11 phenolic acids (hydroxybenzoic and hydroxycinnamic acid derivatives); and eight flavonoids (rhamnetin and quercetin derivatives). The findings for T. Atkins skin constitute the first report of such a high number and diversity of compounds. Also, it is the first time that the presence of gallotannin decamers and undecamers are reported in the skin and flesh of Keitt cultivar and in T. Atkins skins. In addition, total phenolic content (TPC) was measured with high values especially for fruits' skins, with a TPC of 698.65 and 644.17 mg gallic acid equivalents/g extract, respectively, for Keitt and T. Atkins cultivars. Antioxidant potential using 2,2-diphenyl-1-picrylhidrazyl (DPPH) and oxygen radical absorbance capacity (ORAC) methods were evaluated, with T. Atkins skin showing the best values for both DPPH (IC_{50} = 9.97 µg/mL) and ORAC (11.02 mmol TE/g extract). A significant negative correlation was found for samples between TPC and DPPH antioxidant values ($r = -0.960$, $p < 0.05$), as well as a significant positive correlation between TPC and ORAC ($r = 0.910$, $p < 0.05$) and between DPPH and ORAC antioxidant methods ($r = 0.989$, $p < 0.05$). Also, cytotoxicity was evaluated in gastric adenocarcinoma (AGS), hepatocarcinoma (HepG2), and colon adenocarcinoma (SW620), with T. Atkins skin showing the best results (IC_{50} = 138–175 µg/mL). Finally, for AGS and SW 620 cell lines particularly, a high significant negative correlation was found between cytotoxic activity and gallotannins ($r = -0.977$ and $r = -0.940$, respectively) while for the HepG2 cell line, the highest significant negative correlation was found with xanthonoids compounds ($r = -0.921$).

Keywords: *Mangifera indica*; *mango*; UPLC; ESI-MS; polyphenols; xanthonoids; gallotannins; hydroxybenzophenones; mass spectrometry; antioxidant; antitumoral

---

1. Introduction

Several studies have linked vegetable consumption, especially fruits, with a reduced risk for cardiovascular disease and cancer, thus the importance of metabolites' characterization. Mango (*Mangifera indica* L.) is a commercial fruit cultivated worldwide that holds the fifth position in total production amongst the main fruit crops, with 5.4 million hectares in approximately 100 countries, especially in areas with subtropical and tropical climates [1]. Out of the large number of cultivars reported, Keitt and Tommy Atkins are the most important commercialized mango cultivars in Costa Rica.

The bioactive effects reported for *M. indica* include antioxidant activity, anti-inflammatory, antipyretic, antibacterial, antiviral, antimicrobial, and anticancer, as well as hepatoprotective and gastroprotective properties, in addition to immunomodulatory and lipid-lowering drug effects [2–5]. Particularly, it has been reported that mango exhibits antiproliferative activity in MDA-MB-231 adenocarcinoma breast cell lines, HepG2 liver, and HL-60 leukemia cancer cells [6], as well as antitumoral effects on MCF-7 breast carcinoma cells [7], Molt-4 leukemia, A-549 lung, LnCap prostate, and SW-480 colon cancer cells [8]. These studies report different results depending on mango cultivar and on cancer cell lines; however, the effects have mainly been attributed to the fruits' polyphenolic contents.

In fact, several studies have reported polyphenolics benefits on health based on findings from the above bioactivities [9] and have established their role in reducing the risk of degenerative and chronic diseases, therefore contributing to long-term health protection [10]. For instance, their contents in fruits have been associated to a lower risk for cardiovascular diseases and cancer, hence the increase in interest in fruit consumption and the importance of scientific research for polyphenols' characterization and their associated valuable effects on health.

Previous studies of *M. indica* polyphenols have focused, for instance, on properties of xanthonoid compounds, mainly mangiferin isomers [11,12] and gallotannins [13], which have been studied for their anticarcinogenic effects [6,14]. Other studies have involved gallotannins and hydroxybenzophenones [15], xanthonoids and flavonoids [16,17], or phenolic acids and gallotannins [18]. Few reports have studied all five types of compounds [19,20] and their biological activities [21].

Polyphenols' antioxidant properties have been linked, among others, with their anti-inflammatory and anticancer activities, which have been reported to increase with gallotannins' degree of polymerization of specific structures, which has been found to enhance such properties [22], thus further knowledge on phenolic structures' characterization in mango fruits would contribute to a better understanding of their implications in the fruits' quality as a source of dietary compounds with potential biological properties.

Therefore, the objective of the present work was to obtain enriched polyphenolic extracts of fruits from *M. indica* commercial cultivars in Costa Rica and to characterize them through ultra performance liquid chromatography coupled with high resolution mass spectrometry (UPLC-DAD-ESI-MS), with an emphasis on the five types of compounds previously reported. An evaluation of the total polyphenolic contents and antioxidant activity using 2,2-diphenyl-1-picrylhidrazyl (DPPH) and oxygen radical absorbance capacity (ORAC) methods, as well as the cytotoxic activity (MTT) on the adenocarcinoma AGS gastric cell line, adenocarcinoma HepG2 hepatic cell line, and adenocarcinoma SW620 colon cell line was also carried out in the different extracts.

2. Materials and Methods

2.1. Materials, Reagents and Solvents

Mangifera indica fruits were acquired in the ripe state from a local producer from Marichal Orotina Orotina (Keitt cultivar) and Fabio Baudrit Station (Alajuela). Cultivars were confirmed with the support of the Costa Rican National Herbarium and vouchers are deposited there. Reagents, such as fluorescein, 2,2-azobis(2-amidinopropane) dihydrochloride (AAPH), 2,2-diphenyl-1-picrylhidrazyl (DPPH), Trolox, gallic acid, Amberlite XAD-7 resin, fetal bovine serum, glutamine, penicillin, streptomycin, amphotericin B, and trypsin–ethylenediaminetetraacetic acid (EDTA), were provided by Sigma-Aldrich (St. Louis, MO, USA). Human gastric adenocarcinoma cell line AGS, human colorectal adenocarcinoma SW 620, and human hepatocellular carcinoma Hep-G2 were obtained from American Type Culture Collection (ATCC, Rockville, MD, USA). In order to evaluate the specificity of the cytotoxic activity towards these cancer cells with respect to normal cells, a selectivity index (SI) was determined by also measuring the cytotoxicity on normal non-cancer cells, according to previous publications [23–25]. Different cell lines are used in the literature, such as normal mouse subcutaneous fibroblast L929 in studies evaluating cytotoxicity on HeLa and SiHa cervical cancer cells [26]; normal human dermal fibroblast TelCOFSO2MA used in comparative cytotoxicity studies with Caco-2 colon and OE19 esophageal adenocarcinoma cell lines [27] and normal monkey epithelial kidney Vero cells used as non-tumoral control cells in studies evaluating cytotoxicity towards MCF-7 breast and HeLa cervix cancer cells [24]; Caco-2 colon and A549 lung cancer cells [23]; AGS gastric and SW620 adenocarcinoma cells [28]; and malignant HepG2 hepatoma cells [25]. These Vero cell lines were selected for this study due to previous reports and accessibility (American Type Culture Collection, Rockville, MD, USA). Finally, solvents, such as acetone, chloroform, and methanol, were purchased from Baker (Center Valley, PA, USA), while DMSO was acquired from Sigma-Aldrich (St. Louis, MO, USA).

2.2. Phenolic Extracts from Mangifera. Indica Fruits

M. indica fruits were rinsed in water, peeled, and both the skin and flesh material were freeze-dried in a Free Zone at −105 °C, 4.5 L, Cascade Benchtop Freeze Dry System (Labconco, Kansas, MO, USA). The freeze-dried material was preserved at −20 °C until extraction. Freeze-dried samples were extracted in a Dionex™ ASE™ 150 Accelerated Solvent Extractor (Thermo Scientific™, Walthman, MA, USA) using methanol:water (70:30) as solvent for 7.5 g of sample in a 34 mL cell, at 40 °C. Next, the extract was evaporated under vacuum to eliminate the methanol and the aqueous phase was washed with ethyl acetate and chloroform to remove less-polar compounds. Afterwards, the aqueous extract was evaporated under vacuum to eliminate organic solvent residues and was eluted (2 mL/min) in an Amberlite XAD7 column (150 mm × 20 mm), starting with 300 mL of water to remove sugars, and then with 200 mL each of methanol:water (80:20) and pure methanol to obtain the polyphenols. Finally, the enriched extract was obtained after evaporation to dryness using a Buchi™ 215 (Flawil, Switzerland) rotavapor.

2.3. Total Phenolic Content

The polyphenolic content was determined by modification of the Folin–Ciocalteu (FC) method [29], which is based on the oxidation of the hydroxyl groups of phenols by the mixture of phosphotungstic and phosphomolybdic acids. Briefly, each polyphenolic enriched extract was dissolved in MeOH (0.1% HCl) to obtain a 500 ppm solution and 2 mL were combined with 0.5 mL of FC reagent. Afterwards 10 mL of Na_2CO_3 (7.5%) were added and the volume was completed to 25 mL with water. Blanks were prepared in a similar way but using 0.5 mL of MeOH (0.1% HCl) instead of the sample. The mixture was left standing in the dark for 1 h and then absorbance was measured at 750 nm. Values obtained were extrapolated in a gallic acid calibration curve. Total phenolic content was expressed as mg gallic acid equivalents (GAE)/g sample. Analyses were performed in triplicate.

2.4. UPLC-DAD-ESI-TQ-MS Analysis

The UPLC-MS system used to analyze the composition of *M. indica* fruit extracts consisted of an LTQ Orbitrap XL mass spectrometer with an Accela 1250 binary Pump, PAL HTC Accela TMO autosampler, PDA detector (Thermo Fisher Scientific, San Jose, CA, USA), and G1316A column compartment (Agilent, Palo Alto, CA, USA). Separation was carried out by a modification of a method previously described [30]. Briefly, a Hypersil Gold AQ RP-C18 UHPLC column (200 mm × 2.1 mm i.d., 1.9 μm, Thermo Fisher Scientific) with an UltraShield pre-column filter (Analytical Scientific Instruments, Richmond, CA, USA) were used at a flow rate of 0.3 mL/min. Mobile phases A and B consist of a combination of 0.1% formic acid in water, *v/v* and 0.1% formic acid in acetonitrile, *v/v*, respectively. The linear gradient is from 4% to 20% B (*v/v*) at 20 min, to 35% B at 30 min and to 100% B at 31 min, and held at 100% B to 35 min. The UV/Vis spectra were acquired from 200 to 700 nm. The mass spectrometer was calibrated using Pierce™ LTQ ESI Negative Ion Calibration Solution, and the conditions for the negative electrospray ionization mode used were set as follows: Sheath gas, 70 (arbitrary units); aux and sweep gas, 15 (arbitrary units); spray voltage, 4.8 kV; capillary temperature, 300 °C; capillary voltage, 15 V; and tube lens, 70 V. The mass range was from 100 to 2000 amu with a resolution of 30,000, FTMS AGC target at 2e5, FT-MS/MS AGC target at 1×10^5, isolation width of 1.5 *amu*, and max ion injection time of 500 ms. A clean chromatographic separation was obtained, and the most intense ion was selected for the data-dependent scan to offer MS^2 to MS^5 product ions, respectively, with a normalization collision energy at 35%.

2.5. DPPH Radical-Scavenging Activity

DPPH evaluation was performed by a modification of the original method [31], based on antioxidant determinations using a stable free radical. Briefly, a solution of 2,2-diphenyl-1-picrylhidrazyl (DPPH) (0.25 mM) was prepared using methanol as the solvent. Next, 0.5 mL of this solution were mixed with 1 mL of each polyphenolic-enriched extract at different concentrations ranging between 4 and 40 ppm, and incubated at 25 °C in the dark for 30 min. The mixture absorbance was measured at 517 nm. Blanks were prepared for each concentration. The percentage of the radical-scavenging activity of the sample was plotted against its concentration to calculate IC_{50} (μg/mL). The samples were analyzed in three independent assays. Results were expressed as IC_{50} (μg/mL), which is the amount of sample required to reach 50% radical-scavenging activity.

2.6. ORAC Antioxidant Activity

The ORAC (oxygen radical absorbance capacity) antioxidant activity was determined by modification of a method using fluorescein as a fluorescence probe [32]. Briefly, the reaction was performed in 75 mM phosphate buffer (pH 7.4) at 37 °C. The final assay mixture consisted of AAPH (12 mM), fluorescein (70 nM), and either Trolox (1–8 μM) or the extract at different concentrations. Fluorescence was recorded every minute for 98 min in black 96-well untreated microplates (Nunc, Denmark), using a Polarstar Galaxy plate reader (BMG Labtechnologies GmbH, Offenburg, Germany) with 485-P excitation and 520-P emission filters. Fluostar Galaxy software version 4.11–0 (BMG Labtechnologies GmbH, Offenburg, Germany) was used to measure fluorescence. Fluorescein was diluted from a stock solution (1.17 mM) in 75 mM phosphate buffer (pH 7.4), while AAPH and Trolox solutions were freshly prepared. All reaction mixtures were prepared in duplicate and three independent runs were completed for each extract. Fluorescence measurements were normalized to the curve of the blank (no antioxidant). From the normalized curves, the area under the fluorescence decay curve (AUC) was calculated as:

$$AUC = 1 + \sum_{i=1}^{i=98} \int_i \! / \int_0, \tag{1}$$

where $\int_0$ is the initial fluorescence reading at 0 min and $\int_i$ is the fluorescence reading at time *i*. The net AUC corresponding to a sample was calculated as follows:

$$\text{Net AUC} = AUC_{\text{antioxidant}} - AUC_{\text{blank}}. \tag{2}$$

The regression equation between the net AUC and the antioxidant concentration was calculated. The ORAC value was estimated by dividing the slope of the latter equation by the slope of the Trolox line obtained for the same assay. Final ORAC values are expressed as mmol of Trolox equivalents (TE)/g of phenolic extract.

2.7. Evaluation of Cytotoxicity of Extracts

2.7.1. Cell Culture

The human gastric adenocarcinoma cell line AGS, human colorectal adenocarcinoma SW 620, human hepatocellular carcinoma Hep-G2, and monkey normal epithelial kidney cells Vero were grown in minimum essential Eagle's medium (MEM) containing 10% fetal bovine serum (FBS) in the presence of 2 mmol/L glutamine, 100 IU mL^{-1} penicillin, 100 µg/mL streptomycin, and 0.25 µg/mL amphotericin B. The cells were grown in a humidified atmosphere containing 5% CO_2 at 37 °C and sub-cultured by detaching with trypsin–EDTA solution at about 70% to 80% confluence. For the experiments, 100 µL of a cell suspension of 1.5×10^5 cells/mL were seeded overnight into 96-well plates. The cells were further exposed for 48 h to various concentrations of extracts (50 µL), dissolved in DMSO, and diluted with cell culture medium to final concentrations between 15 and 500 µg/mL. The DMSO concentrations used in the experiments were below 0.1% (*v/v*) and control cultures were prepared with the addition of DMSO (vehicle control).

2.7.2. Assessment of Cytotoxicity by MTT Assay

After incubation for 48 h, MTT assays were performed to evaluate cytotoxicity. Briefly, the medium was eliminated, cells were washed twice with 100 µL of PBS, and incubated with 100 µL MTT solution (3-(4,5-dimethylthiazolyl-2)-2,5-diphenyltetrazolium bromide, 5 mg/mL in cell culture medium) for 2 h at 37 °C. The formazan crystals formed were dissolved in 100 µL of ethanol 95% and the absorbance was read at 570 nm in a microplate reader. Dose–response curves were established for each extract and the concentration that is enough to reduce the cell viability by 50% (IC_{50}) was calculated.

In order to evaluate if the cytotoxicity activity was specific against the cancer cells, a selectivity index (SI) was determined. This index is defined as the ratio of IC_{50} values of normal epithelial kidney cells (Vero) to cancer cells (AGS, HepG2, or SW620).

2.8. Statistical Analysis

In order to evaluate if the total phenolic contents (TPC) contribute to the antioxidant activity evaluated with the DPPH and ORAC methodologies, a correlation analysis was carried out between the TPC values and the DPPH and ORAC results. Also, one-way analysis of variance (ANOVA) followed by Tukey's post hoc test was applied to the TPC, DPPH, and ORAC values, and differences were considered significant at $p < 0.05$.

3. Results and Discussion

3.1. Phenolic Yield and Total Phenolic Contents

The extraction process described in the Materials and Methods section allowed the phenolic-enriched extracts to be obtained, which are expressed as g of phenolic enriched extract/100 g of dry material and are summarized in Table 1. Keitt cultivar skin presented the highest yield (2.77 g extract/100 g dry material) whereas Tommy Atkins flesh showed the lowest value (0.57 g extract/100 g dry material). In both cultivars, skin extract yields were higher than flesh extracts.

Table 1. Extraction yield and total phenolic content.

Sample	Lyophilization Yield (g/100 g) [1]	Extraction Yield (g/100 g) [2]	Total Phenolic Content (TPC) (mg/g) [3,4,5]
Keitt			
Skin	20.8	2.77	698.65 ± 0.47 [a]
Flesh	16.9	0.69	291.14 ± 1.19 [b]
T. Atkins			
Skin	21.5	2.75	644.17 ± 5.79 [c]
Flesh	17.6	0.57	162.67 ± 1.46 [d]

[1] g of dry material/100 g of fresh weight [2] g of phenolic enriched extract/100 g of dry material [3] mg of gallic acid equivalents (GAE)/g extract. [4] Values are expressed as mean ± standard deviation (S.D.) [5] Different superscript letters in the column indicate differences are significant at $p < 0.05$.

The total phenolic contents (TPC) summarized in Table 1 show results ranging between 162.7 and 698.7 gallic acid equivalents (GAE)/g dry extract. The one-way analysis of variance (ANOVA), with a Tukey post hoc as the statistical test, showed a significant difference ($p < 0.05$) between results for the skin and flesh of *M. indica* samples, with a much higher average for skins corresponding to 671.4 GAE/g dry extract compared to a three times lower average value of 226.9 GAE/g dry extract for flesh. Total phenolic contents (TPC) previous reports for the flesh of T. Atkins cultivars from Mexico and Spain range between 15.3 and 21.77 mg GAE/100 g FW [8,33], whereas our finding of 16.3 mg GAE/100 g FW (value calculated using TPC and lyophilization yields from Table 1) fits within that range. Meanwhile, values reported for the skins of T. Atkins (43.17 mg GAE/100 g FW) and Pica (72.01 mg GAE/100 g FW) cultivars from Chile [34] are lower than our result of 380.9 mg GAE/100 g FW (value calculated using TPC and lyophilization yields from Table 1). In respect to Keitt flesh, results from the literature show variability, reporting values ranging between 17.99 and 59.43 mg GAE/100 g FW for Keitt and other cultivars from Italy and China [21,33] whereas our result of 33.9 mg GAE/100 g FW (value calculated using TPC and lyophilization yields from Table 1) fits in that range. Finally, regarding Keitt skin, previous results for this and other cultivars from China range between 368.52 and 641.9 mg GAE/100 g FW [21]. Our finding of 402.5 mg GAE/100 g FW (value calculated using TPC and lyophilization yields from Table 1) fit within that range and are higher than values found for the Irwin cultivar (26.9 mg GAE/g extract) from Korea [35]. TPC content variations are linked to polyphenols present in extracts [36,37] and the influence of these metabolites in extracts' biological properties, such as the antioxidant capacity [38,39] and cytotoxic activity [14,22], as discussed in the following sections.

3.2. Profile by UPLC-DAD-ESI-TQ-MS Analysis

The UPLC-DAD-ESI-MS/MS analysis described in the Materials and Methods section allowed identification of 71 compounds, including 32 gallates and gallotannins, six xanthonoids, eight hydroxybenzophenones, eight flavonoids, and 11 phenolic acids and derivatives, in Costa Rican Keiit and T. Atkins commercial cultivars. Figures 1 and 2 show the chromatograms of the 71 different compounds and Table 2 summarizes the results of the identification analysis.

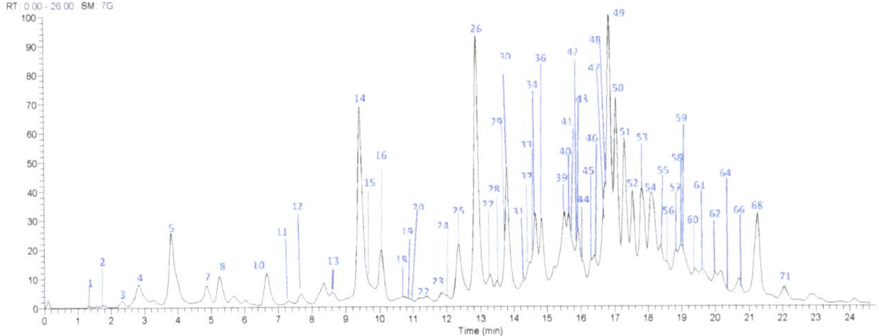

Figure 1. High Performance Liquid Chromatography (HPLC) chromatogram of *Mangifera indica* Tommy Atkins cultivar skin extract, in a Hypersil Gold AQ RP-C18 column (200 mm × 2.1 mm × 1.9 µm) using an LTQ Orbitrap XL Mass spectrometer (Thermo Scientific™, Walthman, MA, USA) in a mass range from 100 to 2000 amu.

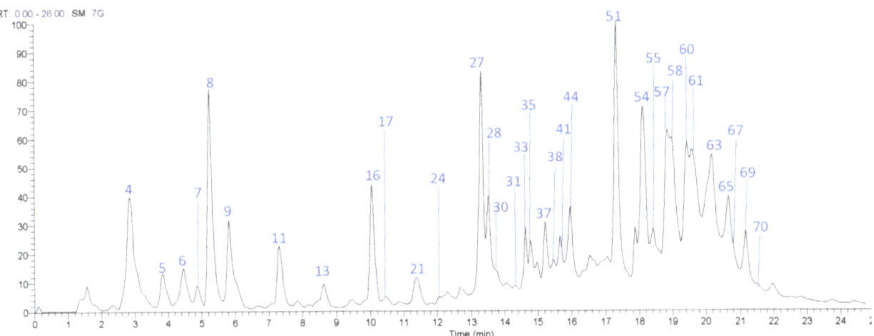

Figure 2. High Performance Liquid Chromatography (HPLC) chromatogram of *Mangifera indica* Keitt flesh extract in a Hypersil Gold AQ RP-C18 column (200 mm × 2.1 mm × 1.9 µm) using an LTQ Orbitrap XL mass spectrometer (Thermo Scientific™, Walthman, MA, USA) in a mass range from 100 to 2000 amu.

Table 2. Profile of phenolic compounds identified by UPLC-DAD-ESI-TQ-MS analysis for mangoes Keitt and T. Atkins samples.

Peak	Identification	Rt (min)	$[M-H]^-$	Molecular Formula	Error (ppm)	MS2 Fragments	MS3 Fragments	Keith Skin	Keit Flesh	T. A. Skin	T. A. Flesh
						Phenolic Acis					
6	Hydroxybenzoic acid hexoside (isomer I of II)	4.43	299.078	$C_{13}H_{15}O_8$	6.31	[299]: 137(100)	[299→137]: 93(100)		x		
9	Hydroxybenzoic acid hexoside (isomer II of II)	5.80	299.0775	$C_{13}H_{15}O_8$	4.37	[299]: 137(100). 179(75). 239(79)	[299→137]: 93(100)	x	x		
12	5-hydroxyferuloyl hexoside	7.64	371.0993	$C_{16}H_{19}O_{10}$	5.38	[371]: 209(90). 233(100)	[371→233]: 191(100), 205(89)			x	
17	Ferulic acid	10.47	193.0515	$C_{10}H_9O_4$	10.07	[193]: 149(100). 178(73)	[193→149]: 134(100)		x		
24	Sinapic acid	12.03	223.0618	$C_{11}H_{11}O_5$	7.40	[223]: 164(17). 179(32). 208(100)	[223→208]: 164(100)		x	x	
27	Sinapic acid O-pentosyl-hexoside	13.27	517.2304	$C_{24}H_{37}O_{12}$	4.75	[517]: 205(93). 385(100)	[517→385]: 153(100), 205(87), 223(90)	x	x	x	x
28	Dihydrosinapic acid O-pentosyl-hexoside	13.47	519.2462	$C_{24}H_{39}O_{12}$	5.04	[519]: 387(100)	[519→387]: 161(100), 225(63)	x	x	x	x
35	Syringic acid hexoside derivative (isomer I of III)	14.75	403.1621	$C_{18}H_{27}O_{10}$	5.55	[403]: 241(100)	[403→241]: 197(100)		x		
37	Syringic acid hexoside derivative (isomer II of III)	15.18	403.1621	$C_{18}H_{27}O_{10}$	5.62	[403]: 241(100)	[403→241]: 197(100)	x	x		
38	Syringic acid hexoside derivative (isomer III of III)	15.47	403.1618	$C_{18}H_{27}O_{10}$	4.70	[403]: 241(100)	[403→241]: 197(100)		x		
47	Ellagic acid	16.55	300.9999	$C_{14}H_5O_8$	6.60	[301]: 229(52). 257(100)		x		x	x
						Other acids					
1	Quinic acid	1.58	191.0568	$C_7H_{11}O_6$	9.24	[191]: 85(69). 93(57). 127(100). 173(83)		x		x	
16	Dihydrophaseic acid hexoside (isomer I of II)	9.99	443.1934	$C_{21}H_{31}O_{10}$	5.00	[443]: 161(90). 189(48). 219(83). 237(100). 281(44). 425(35)	[443→237]: 219(100)	x	x	x	x
33	Dihydrophaseic acid hexoside (isomer II of II)	14.59	443.1940	$C_{21}H_{31}O_{10}$	6.45	[443]: 189(51). 219(52). 237(87). 263(100). 281(34). 399(33). 425(97)		x	x	x	x

Table 2. *Cont.*

Peak	Identification	Rt (min)	[M−H]⁻	Molecular Formula	Error (ppm)	MS2 Fragments	MS3 Fragments	Keith Skin	Keit Flesh	T. A. Skin	T. A. Flesh
						Gallates and Gallotannins					
2	Galloyl dihexoside (isomer I of III)	1.79	493.1210	$C_{19}H_{25}O_{15}$	4.47	[493]: 313(100)	[493→313]: 169(100), 223(52)			x	
3	Galloyl dihexoside (isomer II of III)	2.35	493.1213	$C_{19}H_{25}O_{15}$	4.98	[493]: 313(100)	[493→313]: 169(100), 223(52)			x	
4	Galloyl *O*-hexose (isomer I of II)	2.67	331.0671	$C_{13}H_{15}O_{10}$	3.34	[331]: 169(100), 211(31), 271(79)	[331→169]: 125(100)	x	x	x	x
5	Galloylquinic acid	3.81	343.0667	$C_{14}H_{15}O_{10}$	2.24	[343]: 191(100)	[343→191]: 85(77), 93(61), 111(35), 126(100), 173(79)	x	x	x	
7	Galloyl *O*-dihexoside (isomer III of III)	4.86	493.1214	$C_{19}H_{25}O_{15}$	3.09	[493]: 313(100)	[493→313]: 169(100), 223(52)	x	x	x	
8	Galloyl *O*-hexose (isomer II of II)	5.23	331.0674	$C_{13}H_{15}O_{10}$	4.43	[331]: 169(100)	[331→169]: 125(100)	x	x	x	
11	Di-*O*-galloyl hexose (isomer I of II)	7.28	483.0799	$C_{20}H_{19}O_{14}$	6.21	[483]: 169(100)	[483→169]: 125(100)	x	x	x	
13	Methyl-gallate isomer	8.63	357.0834	$C_{15}H_{17}O_{10}$	4.89	[357]: 169(100)			x	x	
15	Di-*O*-galloyl quinic acid	9.74	495.0794	$C_{21}H_{19}O_{14}$	4.28	[495]: 343(100)	[495→343]: 169(100)	x		x	
19	Tri-*O*-galloyl hexose (isomer I of III)	10.82	635.08978	$C_{27}H_{23}O_{18}$	2.98	[635]: 465(100), 483(95)	[635→465]: 168(59), 295(31), 313(87), 421(100)	x		x	
20	Di-*O*-galloyl hexose (isomer II of II)	10.94	483.0792	$C_{20}H_{19}O_{14}$	2.26	[483]: 331(100)	[483→331]: 169(100)	x		x	
21	Hydroxybenzoyl galloyl hexoside	11.35	451.0900	$C_{20}H_{19}O_{12}$	6.42	[451]: 313(100)	[451→313]: 169(100)		x		
22	Tri-*O*-galloyl hexose (isomer II of III)	11.38	635.0888	$C_{27}H_{23}O_{18}$	1.35	[635]: 465(100), 483(95)	[635→465]: 168(59), 295(31), 313(87), 421(100)	x		x	
29	Tri-*O*-galloyl hexose (isomer III of III)	13.68	635.08942	$C_{27}H_{23}O_{18}$	2.41	[635]: 465(100), 483(95)	[635→465]: 168(59), 295(31), 313(87), 421(100)	x		x	
31	Tetra-*O*-galloyl hexose (isomer I of VI)	14.23	787.1008	$C_{34}H_{27}O_{22}$	1.98	[787]: 635(100)	[787→635]: 423(77), 465(100), 483(99)	x	x	x	

Table 2. *Cont.*

Peak	Identification	Rt (min)	[M−H]⁻	Molecular Formula	Error (ppm)	MS2 Fragments	MS3 Fragments	Keith Skin	Keit Flesh	T. A. Skin	T. A. Flesh
32	Tetra-*O*-galloyl hexose (isomer II of VI)	14.37	787.1013	$C_{34}H_{27}O_{22}$	3.06	[787]: 635(100)	[787→635]: 423(77), 465(100), 483(99)	x		x	
40	Tetra-*O*-galloyl hexose (isomer III of VI)	15.61	787.1015	$C_{34}H_{27}O_{22}$	2.65	[787]: 617(100), 635(53)	[787→617]: 447(32), 465(100)	x		x	
41	Tetra-*O*-galloyl hexose (isomer IV of VI)	15.73	787.1017	$C_{34}H_{27}O_{22}$	3.60	[787]: 617(25), 635(100)	[787→635]: 423(74), 465(76), 483(100)	x	x	x	
44	Tetra-*O*-galloyl hexose (isomer V of VI)	15.98	787.1016	$C_{34}H_{27}O_{22}$	3.52	[787]: 617(100)	[787→635]: 403(62), 447(65), 465(100)	x	x	x	
45	Tetra-*O*-galloyl hexose (isomer VI of VI)	16.32	787.1010	$C_{34}H_{27}O_{22}$	2.75	[787]: 635(100)	[787→635]: 423(77), 465(100), 483(99)	x		x	
51	Penta-*O*-galloyl hexose	17.27	939.1132	$C_{41}H_{31}O_{26}$	−2.68	[939]: 769(100)	[939→769]: 599(31), 601(32), 617(100)	x	x	x	
54	Hexa-*O*-galloyl hexose (isomer I of III)	18.07	1091.1238	$C_{48}H_{35}O_{30}$	0.6	[1091]: 939(100)	[1091→939]: 769(100)	x	x	x	
55	Hexa-*O*-galloyl hexose (isomer II of III)	18.37	1091.1227	$C_{48}H_{35}O_{30}$	−0.41	[1091]: 939(100)	[1091→939]: 769(100)	x	x	x	
56	Hexa-*O*-galloyl hexose (isomer III of III)	18.52	1091.1235	$C_{48}H_{35}O_{30}$	0.38	[1091]: 939(100)	[1091→939]: 769(100)	x		x	
57	Hepta-*O*-galloyl hexose (isomer I of III)	18.80	1243.1351	$C_{55}H_{39}O_{34}$	−0.10	[1243]: 939(48), 1091(100)	[1243→1091]: 939(100)	x	x	x	
58	Hepta-*O*-galloyl hexose (isomer II of III)	18.94	1243.1349	$C_{55}H_{39}O_{34}$	−2.18	[1243]: 939(48), 1091(100)	[1243→1091]: 939(100)	x	x	x	
59	Hepta-*O*-galloyl hexose (isomer III of II)	19.06	1243.1352	$C_{55}H_{39}O_{34}$	2.74	[1243]: 939(56), 1091(100)	[1243→1091]: 939(100)	x		x	
60	Octa-*O*-galloyl hexose (isomer I of II)	19.39	1395.1466	$C_{62}H_{43}O_{38}$	2.81	[1395]: 1243(100), 1244(58)		x	x	x	
61	Octa-*O*-galloyl hexose (isomer II of II)	19.60	1395.1464	$C_{62}H_{43}O_{38}$	2.64	[1395]: 1243(100), 1244(41)		x	x	x	
63	Nona-*O*-galloyl hexose	20.11	1547.1576	$C_{69}H_{47}O_{42}$	2.55	[1547]: 1395(100), 1396(62)		x	x		
64	Nona-*O*-galloyl hexose	20.44	1547.1603	$C_{69}H_{47}O_{42}$	4.29	[1547]: 1395(100), 1396(62)		x		x	

Table 2. *Cont.*

Peak	Identification	Rt (min)	$[M-H]^-$	Molecular Formula	Error (ppm)	MS2 Fragments	MS3 Fragments	Keith Skin	Keit Flesh	T. A. Skin	T. A. Flesh
65	Deca-*O*-galloyl hexose	20.60	1699.1690	$C_{76}H_{51}O_{46}$	2.56	[1699]: 1547(100), 1548(53)		x	x		
66	Deca-*O*-galloyl hexose	20.80	1699.1724	$C_{76}H_{51}O_{46}$	4.57	[1699]: 1547(100), 1548(53)		x		x	
67	Undeca-*O*-galloyl hexose (isomer I of III)	20.94	1851.1819	$C_{83}H_{55}O_{50}$	3.42	[1851]: 1547(33), 1699(100), 1700(90)		x	x		
69	Undeca-*O*-galloyl hexose (isomer II of III)	21.28	1851.1805	$C_{83}H_{55}O_{50}$	2.69	[1851]: 1395(43), 1547(47), 1699(100), 1700(81)		x	x		
70	Undeca-*O*-galloyl hexose (isomer III of III)	21.53	1851.1803	$C_{83}H_{55}O_{50}$	2.56	[1851]: 1395(100), 1547(76), 1699(99), 1700(85)		x	x		
						Xanthonoids					
10	Maclurin C-hexoside	6.62	423.0943	$C_{19}H_{19}O_{11}$	5.94	[423]: 303(100)	[423→303]: 193(100)			x	
14	Maclurin 3-C-(2-*O*-galloyl)-hexoside	9.35	575.1047	$C_{26}H_{23}O_{15}$	2.67	[575]: 285(85), 303(100), 313(43), 423(70), 465(31)	[575→303]: 193(100)	x		x	
18	Maclurin-3-C-(2-*O*-hexosyl-galloyl)-hexoside	737.1588		$C_{32}H_{33}O_{20}$	2.78	[737]: 575(100)	[737→575]: 285(89), 303(100), 313(44), 423(80)			x	
25	Iriflophenone 3-C-(2-*O*-galloyl)-hexoside	12.31	559.1101	$C_{26}H_{23}O_{14}$	3.36	[559]: 287(31), 407(100)	[559→407]: 287(100)	x		x	
26	Maclurin 3-C-(2.3-di-*O*-galloyl)-hexoside	12.82	727.1166	$C_{33}H_{27}O_{19}$	3.49	[727]: 575(100)	[727→ 575]: 315(39), 369(38), 405(100), 439(56), 465(37), 485(78)	x		x	
34	Maclurin-3-C-(*p*-hydroxybenzoyl)-hexoside	14.63	543.1149	$C_{26}H_{23}O_{13}$	5.91	[543]: 285(100)	[543→285]: 175(100)			x	
36	Iriflophenone 3-C-(di-*O*-galloyl)-hexoside	14.83	711.1216	$C_{33}H_{27}O_{18}$	3.36	[711]: 559(100)	[711→559]: 389(100)			x	

Table 2. *Cont.*

Peak	Identification	Rt (min)	[M−H]⁻	Molecular Formula	Error (ppm)	MS2 Fragments	MS3 Fragments	Keith Skin	Keit Flesh	T. A. Skin	T. A. Flesh
						Hydroxybenzophenones					
23	Mangiferin *O*-hexoside	11.84	583.1311	$C_{25}H_{27}O_{16}$	2.98	[583]: 463(65), 493(100), 565(29)	[583→493]: 331(100)			x	
30	Manguiferin/Isomangiferin	13.85	421.0787	$C_{19}H_{17}O_{11}$	2.13	[421]: 301(100), 331(94)	[421→301]: 258(100), 273(73)	x	x	x	x
39	Mangiferin/isomanguiferin *O*-gallate (isomer I of II)	15.48	573.0891	$C_{26}H_{21}O_{15}$	2.82	[573]: 421(100)	[573→421]: 301(100), 331(49)	x		x	
42	Mangiferin/isomanguiferin *O*-gallate (isomer II of II)	15.81	573.0894	$C_{26}H_{21}O_{15}$	3.36	[573]: 283(44), 403(54), 421(100)	[573→421]: 301(100), 331(56)			x	
46	Manguiferin/Isomangiferin	16.43	421.0794	$C_{19}H_{17}O_{11}$	6.82	[421]: 301(100), 331(89), 406(58)	[421→301]: 258(100), 273(80)	x		x	
48	Mangiferin-di-*O*-gallate	16.71	725.1004	$C_{33}H_{25}O_{19}$	2.68	[725]: 573(100)	[757→573]: 403(98), 421(100)			x	
						Flavonoids					
43	Quercetin-3-*O*-pentosyl-hexoside	15.92	595.1303	$C_{26}H_{27}O_{16}$	1.49	[595]: 300(100), 301(42)	[595→300]: 255(57), 271(100)			x	
49	Quercetin *O*-hexoside (isomer I of II)	16.82	463.0893	$C_{21}H_{19}O_{12}$	4.66	[463]: 300(30), 301(100)	[463→301]: 151(67), 179(100)	x		x	x
50	Quercetin *O*-hexoside (isomer II of II)	17.04	463.0898	$C_{21}H_{19}O_{12}$	5.85	[463]: 301(100)	[463→301]: 151(72), 179(100)			x	
52	Quercetin 3-*O*-pentoside (isomer I of II)	17.54	433.0786	$C_{20}H_{17}O_{11}$	4.72	[433]: 301(100)	[433→301]: 179(100), 151(80)			x	
53	Quercetin 3-*O*-pentoside (isomer II of II)	17.85	433.0788	$C_{20}H_{17}O_{11}$	5.22	[433]: 301(100)	[433→301]: 179(100), 151(69)			x	
62	Rhamnetin 3-*O*-pentosyl-hexoside	19.96	609.1467	$C_{27}H_{29}O_{16}$	1.72	[609]: 299(24), 314(100), 315(49)	[609→314]: 299(100)			x	
68	Rhamnetin 3-*O*-hexoside	21.25	477.1054	$C_{22}H_{21}O_{12}$	5.51	[477]: 315(100)	[477→315]: 165(100), 193(38), 299(60), 300(44)			x	
71	Rhamnetin	22.05	315.0513	$C_{16}H_{11}O_{7}$	4.26	[315]: 271(100)	[315→271]: 256(100)			x	

3.2.1. Benzoic and Hydroxycinnamic Acids

Peaks 6 and 9 had an [M−H]⁻ ion at m/z 299.0780 ($C_{13}H_{15}O_8$) and a main MS2 fragment at m/z 137 [M−H−162]⁻ corresponding to a loss of hexose (Glc). Thus, those peaks were identified as isomers of hydroxybenzoic acid hexoside [40], as shown in Figure 3.

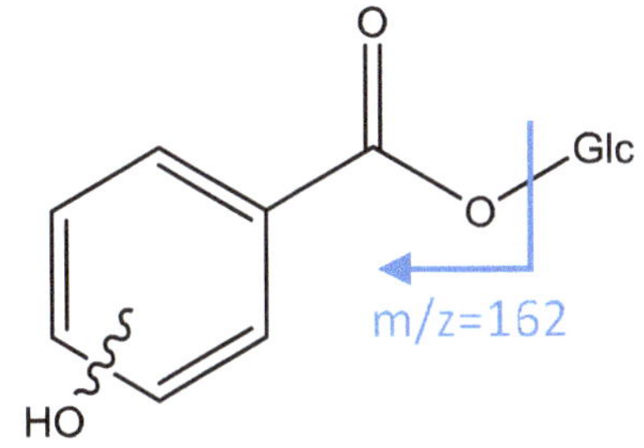

Figure 3. Structure and fragments of acid derivatives (6) and (9).

As represented in Figure 4, peak 12 was identified as 5-hydroxyferuloyl hexoside, with [M−H]⁻ = 371.0993 ($C_{16}H_{19}O_{10}$), with fragments at m/z 233 due to the fragmentation of the aromatic moiety, and at m/z 209 [M−H−162]⁻ due to loss of an hexose. [41] Peak 17 had an [M−H]⁻ ion at 193.0515 ($C_{10}H_9O_4$) that agreed with ferulic acid, with fragments at m/z 178, 149, and 134, due to a loss of methyl groups [M−H−15]⁻, carbon dioxide from the carboxylic acid [M−H−44]⁻, and cleavage through the double bond [M−H−59]⁻ [42]. Peak 24 was assigned to sinapic acid due to its [M−H]⁻ ion at 223.0618 ($C_{11}H_{11}O_5$), which had main fragments at m/z 208 [M−H−15]⁻, 179 [M−H−44]⁻, and 164 [M−H−59]⁻. Peaks 27 ([M−H]⁻ = 517.2304, $C_{24}H_{37}O_{12}$) and 28 ([M−H]⁻ = 519.2462, $C_{24}H_{39}O_{12}$) were identified respectively as sinapic acid O-pentosyl-hexoside and dihydrosinapic acid O-pentosyl-hexoside, which showed an MS2 fragment at [M−H−132]⁻ due to a loss of pentoside, and an MS3 fragment at [M−H−132−162]⁻ due to a subsequent loss of hexoside. [20].

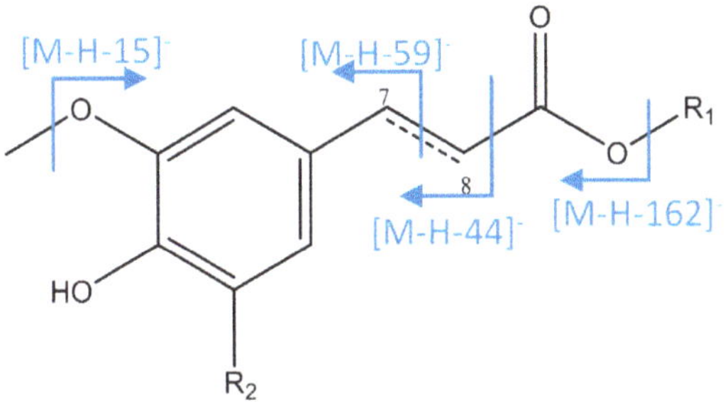

Compound	R₁	R₂	C7–C8
12	-Hexoside	-OH	C=C
17	-H	-H	C=C
24	-H	-OCH₃	C=C
27	-Hexosyl-Pentosyl	-OCH₃	C=C
28	-Hexosyl-Pentosyl	-OCH₃	C-C

Figure 4. Structure and fragments of acid derivatives (12), (17), (24), (27), and (28).

Peaks 35, 37, and 38 (Figure 5) were identified as isomers of a derivate of syringic acid hexoside, which has an [M−H]⁻ ion at 403.1621 ($C_{18}H_{27}O_{10}$) and successive fragments at m/z 241 (loss of hexoside) and 197 (aglycone of syringic acid) [43]. Peak 47 had [M−H]⁻ ion at 300.9995 ($C_{14}H_5O_8$) that is coincident with ellagic acid, whose fragments at m/z 257 and 229 were previously reported [19,44].

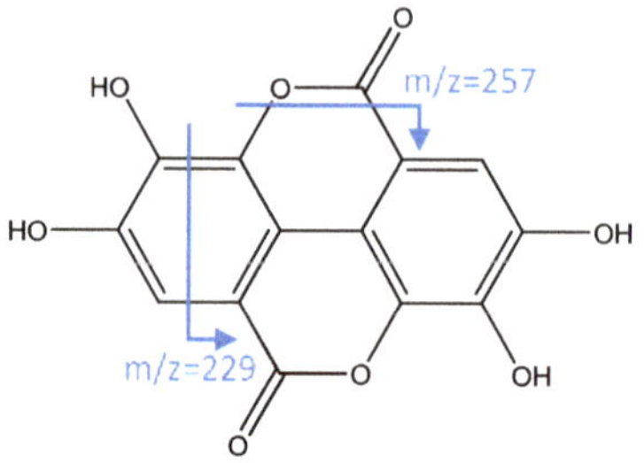

Figure 5. Structure and fragments of acid derivatives (35), (37) and (38).

3.2.2. Other acids

As shown in Figure 6, peak 1, [M−H]⁻ = 191.0568, whose molecular formula was $C_7H_{11}O_6$, agreed with quinic acid. Peaks 16 and 33 had an [M−H]⁻ ion at 443.1934 ($C_{21}H_{31}O_{10}$), and fragments at *m/z* 425 (loss of water), 281 (loss of hexoside), 237 (α-cleavage to carbonyl group), and 219 (cleavage of double bond); so they were identified as isomers of dihydrophseic acid hexoside [45].

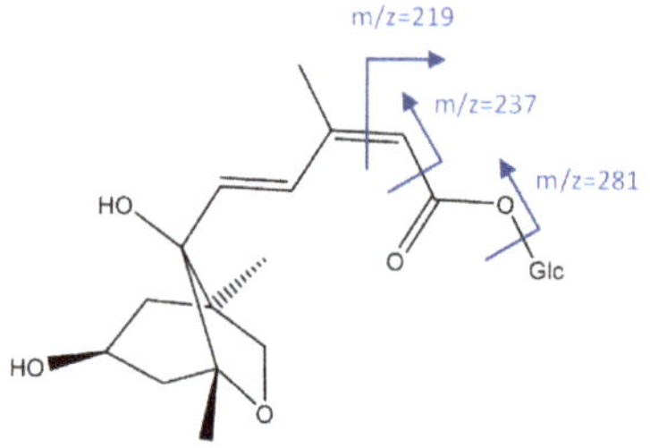

Figure 6. Structure and fragments of acid derivatives (1), (16), and (33).

3.2.3. Gallotanins

As represented in Figure 7, compounds 4 and 8 (galloyl *O*-hexose isomers) show an [M−H]⁻ ion at *m/z* 331.0671 ($C_{13}H_{15}O_{10}$), with a main fragment at *m/z* 169 corresponding to gallic acid. Peak 5 was identified as galloylquinic acid as it had an [M−H]⁻ ion at *m/z* 343.0667 ($C_{14}H_{15}O_{10}$) and a fragment *m/z* 191 consistent with quinic acid due to loss of galloyl [19] Peaks 2, 3, and 7 with an [M−H]⁻ ion at *m/z* 493.1214 ($C_{19}H_{25}O_{15}$) were assigned to galloyl *O*-sucrose due to the MS2 fragment at *m/z* 313, consistent with a loss of an *O*-hexoside, and MS3 fragment at *m/z* 169 corresponding to gallic acid (loss of second hexoside) [46].

Peak 13, [M−H]⁻ 357.0834 ($C_{15}H_{17}O_{10}$), showed a main MS2 fragment at *m/z* 169 and an MS3 fragment at *m/z* 125, which is consistent with gallic acid, thus this peak was tentatively assigned to methyl gallate [19]. Peak 15 (di-*O*-galloyl quinic acid) had an [M−H]⁻ ion at *m/z* 495.0794 ($C_{21}H_{19}O_{14}$), with a main fragment at *m/z* 343 due to the loss of a galloyl. A subsequent loss of quinic acid gives a fragment at *m/z* 169 [47]. Peak 21 was identified as hydroxybenzoyl galloyl hexoside due to its [M−H]⁻ ion at 451.0900 corresponding to a molecular formula of $C_{20}H_{19}O_{12}$. The MS2 fragment at *m/z* 313 occurs due to a loss of hydroxybenzoyl, and MS3 at *m/z* 169 to subsequent loss of hexoside [20].

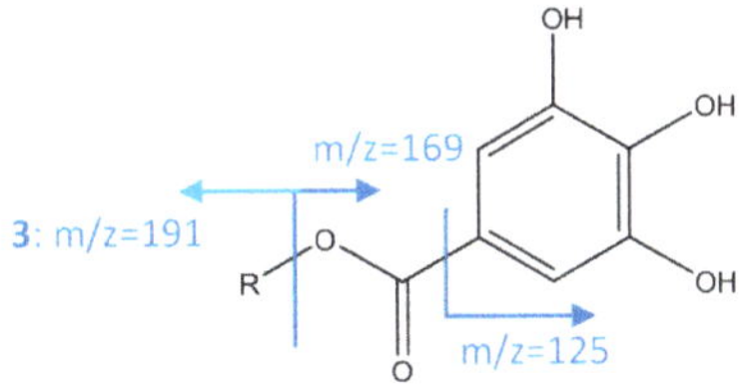

Compound	R
2, 3, 7	-sucrose
4, 8	-hexoside
5	-quinic acid
13	-methyl
15	-galloylquinic acid
21	-hydroxybenzoyl hexoside

Figure 7. Structure and fragments of gallate derivatives.

Peaks 11 and 20 had an $[M-H]^-$ ion at 483.0799 ($C_{20}H_{19}O_{14}$), with a main MS2 fragment at *m/z* 169, consistent with di-*O*-galloyl hexose isomers [19].

A series of peaks (Figure 8) were identified as poly-*O*-galloyl hexoses due to their MS2 fragments at $[M-H-170]^-$ (loss of O-galloyl) and $[M-H-152]^-$ (loss of galloyl), and MS3 fragment at $[M-H-170-152]^-$ due to the loss of a subsequent galloyl. This analysis allowed identification of peaks 19, 22, and 29 ($[M-H]^- = 635.0888$, $C_{27}H_{23}O_{18}$) as tri-*O*-galloyl hexose; peaks 31, 32, 40, 41, 44, and 45 ($[M-H]^- = 787.1008$, $C_{34}H_{27}O_{22}$) as tetra-*O*-galloyl hexose; and peak 51 ($[M-H]^- = 939.1132$, $C_{41}H_{31}O_{26}$) as penta-*O*-galloyl hexose [20].

Compounds with more units of gallate in their structures show an MS2 fragment at $[M-H-152]^-$ due to loss of galloyl, and loss of a second galloyl gave an MS3 fragment at $[M-H-152-152]^-$. Therefore it was possible to identify peaks 54, 55, and 56 ($[M-H]^- = 1091.1238$, $C_{48}H_{35}O_{30}$) as hexa-*O*-galloyl hexose isomers; peaks 57, 58, and 59 ($[M-H]^- = 1234.1351$, $C_{55}H_{39}O_{34}$) as hepta-*O*-galloyl hexose isomers; peaks 60 and 61 ($[M-H]^- = 1395.1466$, $C_{62}H_{43}O_{38}$) as octa-*O*-galloyl hexose isomers; peaks 63 and 64 ($[M-H]^- = 1547.1576$, $C_{69}H_{47}O_{42}$) as nona-*O*-galloyl hexose isomers; peaks 65 and 66 ($[M-H]^- = 1699.1690$, $C_{76}H_{51}O_{46}$) as deca-*O*-galloyl hexose isomers; and peaks 67, 69, and 70 ($[M-H]^- = 1851.1819$, $C_{83}H_{55}O_{50}$) as isomers of undeca-*O*-galloyl hexose isomers [15].

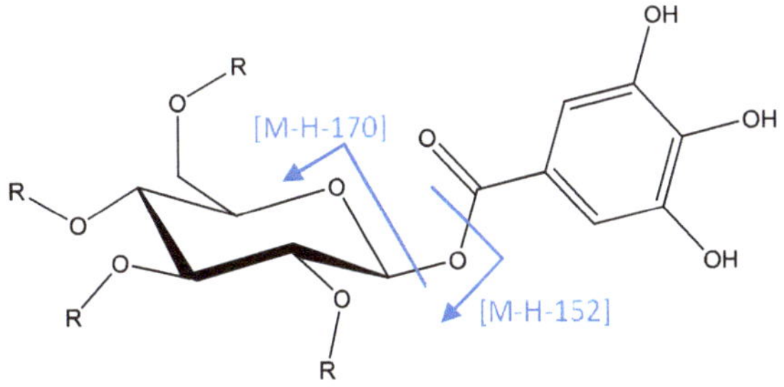

Figure 8. Structure and fragments of gallotannins.

3.2.4. Hydroxybenzophenones

Maclurin 3-C-hexoside derivates (Figure 9) were identified by their molecular formula and characteristic fragment at *m/z* 303 due to retro Diels–Alder cleavage of hexoside. Thus, peak 10

([M−H]$^-$ = 423.0943, C$_{19}$H$_{19}$O$_{11}$) was identified as maclurin 3-C-hexoside. Peak 14 ([M−H]$^-$ = 575.1047, C$_{26}$H$_{23}$O$_{15}$) had main fragments at *m/z* 423 (loss of galloyl) and 303, so it was identified as maclurin 3-C-(2-O-galloyl)-hexoside. Peak 18 ([M−H]$^-$ = 737.1588, C$_{32}$H$_{33}$O$_{20}$) showed a main MS2 fragment at *m/z* 575 due to a loss of hexoside, and MS3 fragments at *m/z* 423 and 303 as the previous peak, so it was identified as maclurin 3-C-(2-O-hexosyl-galloyl)-hexoside. Peak 26 ([M−H]$^-$ = 727.1166, C$_{33}$H$_{27}$O$_{19}$), identified as maclurin 3-C-(2,3-di-O-galloyl)-hexoside, had an MS2 fragment at *m/z* 575 due to the loss of one galloyl, and main MS3 fragments at *m/z* 485 (cleavage of hexoside) and 405 (loss of O-galloyl). Peak 34 ([M−H]$^-$ = 543.1149, C$_{26}$H$_{23}$O$_{13}$) had a main fragment at *m/z* 285 due to the loss of water of a fragment at *m/z* 303, coincident with reports for maclurin-3-C-(p-hydroxybenzoyl)-hexoside [19,48].

Peak 25 ([M−H]$^-$ = 559.1101, C$_{26}$H$_{23}$O$_{14}$) was identified as iriflophenone 3-C-(2-O-galloyl)-hexoside, due to its MS2 fragment at *m/z* 407 ([M−H−152]$^-$, loss of galloyl) and MS3 fragment at *m/z* 287 (hexoside cleavage). Peak 36 ([M−H]$^-$ = 711.1216, C$_{33}$H$_{27}$O$_{18}$) had an MS2 fragment at *m/z* 559 due to a loss of a galloyl, and an MS3 fragment at 387 (loss of O-galloyl), so it was assigned to iriflophenone 3-C-(2,3-di-O-galloyl)-hexoside [49].

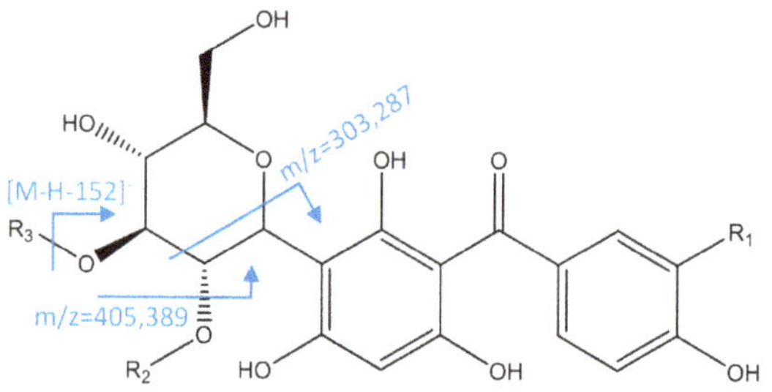

Compound	R$_1$	R$_2$ [a]	R$_3$ [a]
10	-OH	-H	-H
14	-OH	-Galloyl	-H
18	-OH	-Galloyl-Hexosyl	-H
25	-H	-Galloyl	-H
26	-OH	-Galloyl	-Galloyl
34	-OH	p-hydroxybenzoyl	-H
36	-H	-Galloyl	-Galloyl

Figure 9. Structure and fragments of hydroxybenzophenones [a] Positions of substituents may vary among the hexoside ring.

3.2.5. Xanthonoids

As represented in Figure 10, peak 23 had an [M−H]$^-$ ion at 583.1311, with a molecular formula of C$_{25}$H$_{27}$O$_{16}$. It showed MS2 fragments at *m/z* 565 (loss of water), 493 [M−H−90]$^-$, and 463 [M−H−120]$^-$ (both due to cleavage of C-hexoside). The main MS3 fragment at *m/z* 331 [M−H−90-162]$^-$ occurred by the loss of a O-hexoside. Thus, this peak was identified as mangiferin O-hexoside [50].

Peaks 30 and 46 were identified as isomers mangiferin and isomangiferin as they had an [M−H]$^-$ ion at 421.0787 (C$_{19}$H$_{17}$O$_{11}$) and fragments at *m/z* 331 [M−H−90]$^-$ and 301 [M−H−120]$^-$, both of them due to cleavage of glicoside. Peaks 39 and 42 showed an [M−H]$^-$ ion at 573.0891 (C$_{26}$H$_{21}$O$_{15}$), with an MS2 fragment at *m/z* 421 [M−H−152]$^-$ and the same MS3 fragments as mangiferin, thus they were identified as the isomers maguiferin O-gallate and isomangiferin O-gallate [16]. Peak 48 ([M−H]$^-$ = 725.1004, C$_{33}$H$_{25}$O$_{19}$) had an MS2 fragment at *m/z* 573 ([M−H−152]$^-$, loss of galloyl) and MS3 fragments at *m/z* 421 (loss of second galloyl) and 403 (loss of O-galloyl), so it was assigned to mangiferin-di-O-gallate [48].

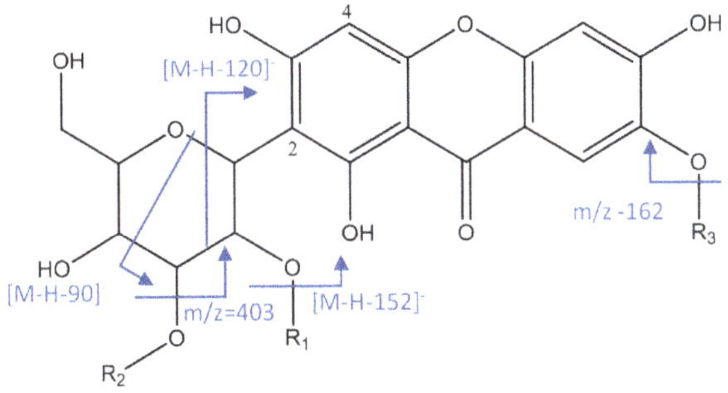

Compound	R₁	R₂	R₃
23	-H	-H	-O-hexoside
30, 46	-H	-H	-H
39, 42	-Galloyl	-H	-H
48	-Galloyl	-Galloyl	-H

Figure 10. Structure and fragments of xanthonoids.

3.2.6. Flavonoids

Peak 43, $[M-H]^- = 595.1303$ ($C_{26}H_{27}O_{16}$) showed (Figure 11) fragments at m/z 300 and 301, coincident with a quercetin aglycone. Thus, this peak was identified as quercetin-3-O-pentosylhexoside. Peaks 49 and 50 were identified as isomers of quercetin 3-O-hexoside, due to their $[M-H]^-$ ion at 463.0893 ($C_{21}H_{19}O_{12}$) and the fragment at m/z 301 ($[M-H-162]^-$, loss of hexoside). Similarly, peaks 52 and 53 ($[M-H]^- = 433.0786$, $C_{20}H_{17}O_{11}$) had the same fragment at m/z 301 $[M-H-132]^-$ due to the loss of a pentoside, so they were identified as isomers of quercetin 3-O-pentoside [19].

Peak 62 had an $[M-H]^-$ ion at 609.1467, consistent with the molecular formula of $C_{27}H_{29}O_{16}$. Its main MS2 fragments at m/z 314 and 315 suggest the presence of a rhamnetin aglycone, so it was assigned to rhamnetin O-pentosylhexoside [51]. Peak 68, with $[M-H]^- = 447.1054$, $C_{22}H_{21}O_{12}$, was identified as rhamnetin O-hexoside, due to its MS2 fragment at m/z 315 (loss of hexoside) [19] while peak 71 ($[M-H]^- = 315.0513$, $C_{16}H_{11}O_7$) was identified as the aglycone rhamnetin, with a main fragment at m/z 271 coincident with previous reports [52,53].

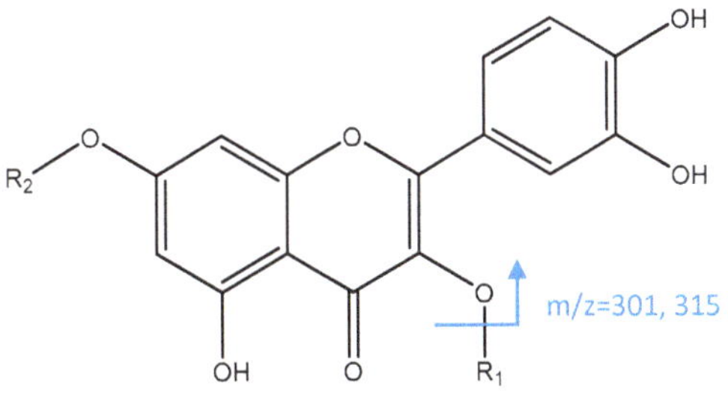

Compound	R₁	R₂
43	-pentosylhexoside	-H
49, 50	-hexoside	-H
52, 53	-pentoside	-H
62	-pentosylhexoside	-CH₃
68	-hexoside	-CH₃
71	-H	-CH₃

Figure 11. Structure and fragments of flavonoids.

Regarding the total number of polyphenols in *M. indica* Keitt and T. Atkins cultivars, 149 compounds were found, comprising 12 xanthonoids, 32 phenolic acids and derivatives, 10 hydroxybenzophenones, 10 flavonoids, and 85 gallates and gallotannins. T. Atkins skins have a greater number of compounds than Keitt skins, which exhibit the highest number of gallates and galloltannins with 32 different compounds. In respect to compound diversity, gallotannins are the most recurrent group of polyphenols in both Keitt and T. Atkins skin samples and in Keitt flesh while phenolic acids are the most abundant group in T. Atkins flesh. Xanthonoids are present in all skin and flesh samples and the other two subfamilies differ in their occurrence. For instance, flavonoids are more abundant in T. Atkins skin while hydroxybenzophenones are found only in the skins of both cultivars.

When comparing reports from the literature, our findings coincide with previous reports on Keitt and T. Atkins cultivars from Brazil and Spain [15,17,20,54] showing that skins have a higher number and diversity of polyphenols than flesh and also that Keitt skin and flesh show a similar phenolic profile while the T. Atkins cultivar shows a different profile, with very low diversity for flesh. On the other hand, our results for the Keitt cultivar, which has been given more attention in the literature, indicate a higher number and diversity of compounds than those reported for cultivars from China, Spain, and the USA [13,19,21], particularly in xanthonoids and gallate and gallotannins, with similar results to other reports on cultivars from Spain [20,54].

On the other hand, when comparing the reports for other cultivars in the literature, our results for Keitt and T. Atkins skin cultivars from Costa Rica show a greater number and diversity of polyphenols than the findings on 18 cultivars from Australia, Kenya, Peru, and Thailand [17], as well as from Brazil [55], China [21], and Spain [19], and similar results to the Sensation cultivar from Spain [54]. In the case of flesh, T. Atkins results are similar to cultivars from Mexico and Haiti [8] and higher than those for the Haden cultivar from Brazil [55], while Keitt flesh shows greater and more diverse results than for the other nine different cultivars from Peru and Thailand [17]; as well as from Brazil [55], Mexico, and Haiti [8]; and similar results to the Sensation cultivar from Spain [54].

Finally, our findings for Costa Rican cultivars indicate an important number of gallotannin oligomers of a higher polymerization degree in both Keitt and T. Atkins skins as well as in Keitt flesh; for instance, galloyl hexose decamers and undecamers that are reported for the first time for these two cultivars. These results are of special interest due to reports showing an enhancement of anti-inflammatory and anticancer activities, with gallotannins' higher degree of polymerization, for instance, showing better results than dimers or the galloyl hexose monomer [22].

3.3. Antioxidant Activity

The DPPH and ORAC values obtained are summarized in Table 3. Regarding DPPH, one-way ANOVA with a Tukey post hoc as the statistical test shows a significant difference ($p < 0.05$) between the results for polyphenolic enriched extracts of skin and flesh. For instance, skins show a higher antioxidant activity, with an average of 10.95 µg extract/mL while the flesh average is 20.15 µg extract/mL. At the individual level, T. Atkins skin presents the greater antioxidant value (IC_{50} = 9.97 µg extract/mL), followed by Keitt skin (IC_{50} = 11.93 µg extract/m). Both are higher than findings for the skin of cultivars from Chile reporting a value of IC_{50} = 46.39 µg extract/mL for T. Atkins cultivar and IC_{50} = 32.23 µg extract/mL for Pica cultivar [34]. This trend is also found when comparing flesh from these cultivars, where IC_{50} = 122.22 µg extract/mL and IC_{50} = 73.76 µg extract/mL were reported for T. Atkins and Pica mangoes [34], lower than our results for Keitt (IC_{50} = 17.78 µg extract/mL) and T. Atkins (IC_{50} = 22.51 µg extract/mL) cultivars. On the other hand, when comparing our DPPH values expressed as mmol TE/g extract (in respect to Trolox IC_{50} = 5.62 µg/mL) with reports in the literature, it is observed that Keitt and T. Atkins skin values fit in the range between α-tocopherol and enju extract (0.42–0.76 mmol TE/g extract), which are existing antioxidant food additives [56].

Table 3. DPPH and ORAC antioxidant activity.

Sample	DPPH [1,2]		ORAC [1,2]
	IC$_{50}$ (µg/mL)	(mmol TE/g Extract)	(mmol TE/g Extract)
Keitt			
Skin	11.93 ± 0.69 [a]	0.47 ± 0.03 [a]	8.30 ± 0.01 [a]
Flesh	17.78 ± 0.33 [b]	0.32 ± 0.01 [b]	5.20 ± 0.12 [b]
T. Atkins			
Skin	9.97 ± 0.36 [c]	0.56 ± 0.02 [c]	11.02 ± 0.11 [c]
Flesh	22.51 ± 0.44 [d]	0.25 ± 0.01 [d]	3.56 ± 0.07 [d]

[1] Values are expressed as mean ± S.D. [2] Different superscript letters in the same column indicate differences are significant at $p < 0.05$. ORAC: oxygen radical absorbance capacity; DPPH: 2,2-diphenyl-1-picrylhidrazyl method.

In respect to ORAC evaluation, the results also show significant differences (ANOVA, $p < 0.05$) between skin and flesh extracts. In fact, skins exhibit a higher antioxidant activity, with an average of 9.66 mmol TE/g extract, while the flesh extract average is 4.68 mmol TE/g extract. At the individual level, T. Atkins skin also shows the highest antioxidant value (11.02 mmol Trolox equivalents/g extract), followed by Keith skin (8.30 mmol TE/g extract), while the lowest value is shown by T. Atkins flesh (3.56 mmol TE/g extract). A previous study on the flesh of different cultivars from Spain reported an ORAC value of 156.6 µmol TE/100 g fresh weight for T. Atkins cultivar and higher values for Francis and Ataulfo cultivars (225.8 and 326.6 µmol TE/100 g fresh weight, respectively) [8], all of which are lower than our results for Keitt flesh (357.1 µmol TE/100 g fresh weight) and T. Atkins flesh (606.4 mmol TE/100 g fresh weight).

Finally, a correlation analysis was carried out between total polyphenolic contents (TPC) values (Table 1) and the DPPH and ORAC results. The findings indicate a significant positive correlation ($p < 0.05$) between both DPPH (mmol TE/g extract) and ORAC antioxidant findings ($r = 0.989$), and between TPC and ORAC values ($r = 0.910$), and a significant negative correlation ($p < 0.05$) was found between the TPC results and DPPH values ($r = -0.960$). Therefore, our results align with previous studies reporting a correlation between the total polyphenolic contents and antioxidant activities of *Mangifera indica* cultivars [54].

3.4. Cytotoxicity

Table 4 and Figure 12 summarize the IC$_{50}$ values for the cytotoxic effect of *M. indica* extracts on different human carcinoma cells, namely AGS, HepG2, and SW620 cell lines, all related to the digestive tract. The development of these types of cancers has been associated to a lower consumption of vegetables and fruits [57], particularly 60% of stomach cancer and 43% of colon cancer are attributed to deficient consumption of vegetables [58]. For this reason, it is interesting to evaluate the cytotoxicity activity of fruit phytochemical-enriched extracts on these tumoral cells.

Table 4. Cytotoxicity of *M. indica* extracts towards gastric (AGS), colon (SW-620), and liver (Hep-G2) carcinoma cells as well as towards Vero non-tumoral cells.

Sample		IC$_{50}$ (µg/mL)			
		AGS [1,2]	SW 620 [1,2]	Hep-G2 [1,2]	Vero [1,2]
Keitt	Skin	197 ± 16 [a,b,*]	223 ± 24 [a,*,&]	309 ± 23 [a,&]	>500 [a,#]
	Flesh	256 ± 32 [a,*]	374 ± 18 [b,&]	369 ± 17 [a,&]	>500 [a,#]
T. Atkins	Skin	138 ± 8 [b,^]	175 ± 7 [a,^]	164 ± 13 [b,^]	278 ± 3 [b,#]
	Flesh	>500 [c,#]	>500 [c,#]	>500 [c,#]	>500 [a,#]

[1] Different superscript letters in the same column indicates that differences are significant at $p < 0.05$. [2] Different superscript signs in the same row indicates differences are significant at $p < 0.05$.

The IC$_{50}$ values shown in Table 4 suggest that a better cytotoxic effect is displayed by the extracts obtained from skin than the extracts obtained from flesh. This difference is statistically significant for all cell lines in the T. Atkins cultivar and for SW-620 cells in the Keitt cultivar. The better results of the skin extracts are in agreement with results previously described for other cultivars [35,59,60]. Also, publications that compared extracts from different parts of the mango highlight the potential of skin and kernel extracts in the cytotoxic activity of different cell lines [61,62].

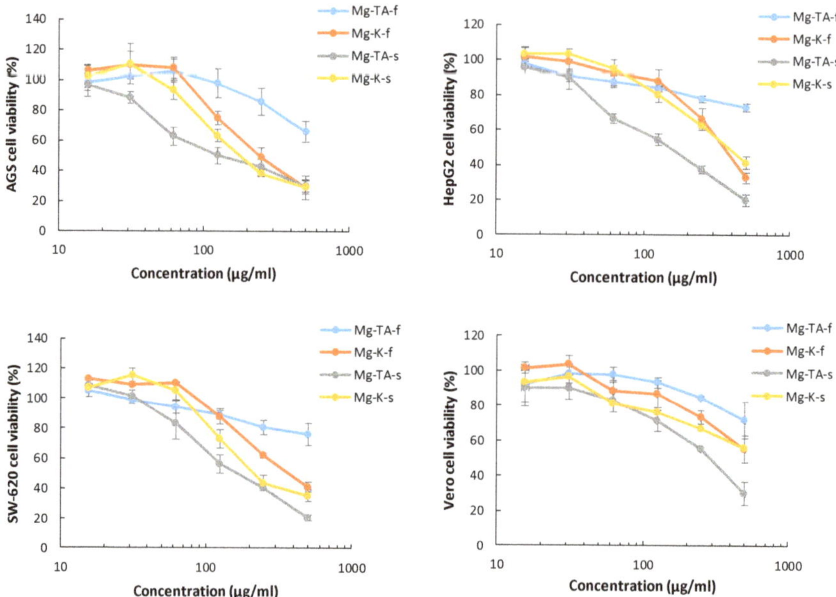

Figure 12. Cytotoxicity dose–response curves of *M. indica* extracts on AGS, HepG2, and SW620 adenocarcinoma cell lines and Vero normal cell lines. Results are presented as mean ± SD of three independent experiments. Mg-TA-f (T. Atkins cultivar flesh), Mg-K-f (Keitt cultivar flesh), Mg-TA-s (T. Atkins cultivar skin), and Mg-K-s (Keitt cultivar skin).

As shown in Table 4, in all assessed samples, the comparison of the cytotoxic effect between the different types of carcinoma cell lines showed better IC$_{50}$ values against the gastric cells (AGS) than the hepatic (HepG2) and colon (SW-620) cell lines. Few reports of cytotoxic effects are available for gastric adenocarcinoma cell lines; for instance, for skin extracts of the Irwi mango cultivar, a dose-dependent effect from 125 to 1000 µg/mL was described on AGS cells [35] and a low cytotoxic effect on Kato-III cells using an ethanolic leaf extract of Okrong mango (IC$_{50}$ > 200 µg/mL) [63]. Our results show better cytotoxic activity in ranges of 15 to 500 µg/mL, with an IC$_{50}$ of 138 ± 8 µg/mL for Tommy Atkins skin and 197 ± 16 µg/mL for Keith skin cultivars.

Table 4 also shows a significantly better cytotoxic effect on the hepatocellular carcinoma cells (HepG2) for the T. Atkins cultivar compared to Keitt mango. In fact, the IC$_{50}$ value assessed for T. Atkins skin is 164 ± 13 µg/mL, which is similar to results reported for the Irwin cultivar, which caused a decrease of 50% viability in concentrations between 125 and 250 µg/mL when incubated with HepG2 cells [35]. However, skin extracts of another mango species, *M. pajang*, showed better cytotoxic activity for HepG2 (IC$_{50}$ = 36.5 µg/mL) than *M. indica* varieties [61].

On the other hand, the cytotoxic effect of *M. indica* cultivar skin extracts on colon carcinoma (SW-620) shows an IC$_{50}$ of 175 ± 7 µg/mL for T. Atkins skin and 223 ± 24 µg/mL for the Keitt cultivar (Table 4). Previous reports for other varieties of *M. indica* cited IC$_{50}$ values over 200 µg/mL for pulp extract [64] in similar colon carcinoma cell lines. In addition to the cytotoxic effect, the extracts used in this study showed a significant difference ($p < 0.05$) between the IC_{50} of the control non-tumoral cell line (Vero) and the tumoral cell lines. This kind of selectivity is a desirable characteristic for any compound with chemotherapy potential [65,66].

Finally, the results of the correlation analysis of cytotoxic activities (Table 4) for each cell line with the abundance of polyphenolic compounds in each sample (Table 2) show a significant negative correlation between cytotoxic effects and the number of polyphenols for all cell lines, AGS ($r = -0.984$), HepG2 ($r = -0.974$), and SW620 ($r = 0.983$), at $p < 0.05$. In fact, T. Atkins skin constitutes the sample with more polyphenolic compounds and with the best cytotoxic activity, followed by the skin of the Keitt cultivar. Specific correlations between the cytotoxic activity and the presence of each type of polyphenolic compounds identified were calculated and it was observed that for AGS and SW 620 cell lines, the cytotoxic activity showed a particularly high significant negative correlation with gallotannins ($r = -0.977$ and $r = -0.940$, respectively), followed by a significant negative correlation of the SW620 cytotoxicity results with xanthonoids ($r = -0.880$). In turn, cytotoxic effects on the HepG2 cell line had the best significant negative correlation with xanthonoids ($r = -0.921$). These specific correlation values between the types of phenolic compounds and cytotoxic activity suggest that gallotannins and xanthonoids play an important role in the toxicity against cancer cells. In fact, previous reports suggest that these two classes of compounds seem to be strong determinants of the anti-tumor activity of mango extracts [14].

One of the main xanthonoids present in the skin extracts of our study is mangiferin and its isomeric forms (Table 2). These compounds have been described in the literature as promising anticancer polyphenols [11,67,68]. The ability of mangiferin to inhibit cancer cells is achieved through several molecular targets, however, one of the important mechanisms is associated with induction of apoptosis [68,69]. Although, besides apoptosis induction, other mechanisms of cell cytotoxicity have been postulated for mangiferin [11], such as cell cycle arrest [70] and a decrease in matrix metalloproteinase activities and reversal of the epithelial–mesenchymal transition [71].

Concerning gallates or gallotannins identified in mango extracts, in vitro studies have shown strong cytotoxic activity; for instance, inhibitions of 55% to 75% in the proliferation of breast, liver, and leukemia cancer cell lines treated with 40 to 80 µg/mL of extracts from Chinese cultivars [6]. The main compounds detected in these mango extracts correspond to pentagalloyl hexose to nonagalloyl hexose isomers. Also, these authors tested penta-galloyl hexoside and gallic acid to treat the same cell lines and confirm the potential role of these compounds as antiproliferatives. Another study on Keitt extract containing galloyl hexosides ranging from pentagalloyl hexose to nonagalloyl hexose inhibited 90% of breast cancer cell lines, with a concentration of 10 µg/mL evidencing strong activity [19].

Table 2 shows that skin extracts from mango contain a high number of gallotannins with different degrees of polymerization, ranging from monogalloyl hexose to undeca-galloyl hexoses. Studies of gallotannins from red maple species in colon and breast cancer cells showed an association between the higher number of galloyl groups in the gallotannins and better anticancer activity [22]. Particularly, the penta galloyl glucoside from fruits, such as maple, gallnuts, and oak, and medicinal herbs (*Galla rhois, Rhus chineensis, Paeonia suffruticosa*) has been widely associated to anticancer effects in prostate, breast, glioma, hepatocellular, and colorectal carcinoma [72–74]. The suggested mechanisms associated to the cytotoxicity involve induction of apoptosis through an increase of Bax/Bcl-2 protein levels, cell cycle arrest in S-phase, and the inhibition of NF-κB activation, with the consequent downregulation of inflammatory cytokines [73].

Even though our results and the cited previous reports suggest that gallates and xanthonoids seem to be strong determinants of the anti-tumor activity, the mechanism of cell cytotoxicity has to be determined and the target molecules of the *M. indica* extract remain to be identified. Also, despite our results indicating a correlation with the presence of xanthonoids and gallates and the cytotoxic activity in tumoral cell lines, these compounds are probably not the only factors responsible for the observed biological effects. The contribution of other components present in the extracts should be clarified because reports in the literature of similar anti-cancer properties are available for phenolic acids and flavonoids [75,76] and hydroxybenzophenones [77].

4. Conclusions

The analysis of phenolic-enriched extracts of Keitt and T. Atkins, the main commercial cultivars of *M. indica* in Costa Rica, using UPLC-DAD-ESI-MS techniques identified a total of 149 compounds, 82 of them in the Keitt cultivar, including 54 different gallotannins, which demonstrates the potential value of this fruit with a greater and more diverse number of compounds than cultivars from different countries and similar to previous important results reported for Keitt fruits from Spain [20,50]. Besides, our results for T. Atkins skin showing 59 compounds, including 30 gallotannins, constitutes the first report of such a high number and diversity of polyphenolic compounds for this cultivar. On the other hand, the TPC, DPPH, and ORAC antioxidant capacity showed high significant correlations ($p < 0.05$), with Keitt and T. Atkins skins exhibiting the highest values. Further, cytotoxicity results were also better for skin extracts in all three adenocarcinoma cell lines studied. For AGS and SW 620 cell lines, cytotoxicity activity showed a particularly high significant negative correlation with gallotannins ($r = -0.977$ and $r = -0.940$, respectively), while for the HepG2 cell line, the highest significant negative correlation was found with xanthonoid compounds ($r = -0.921$). These results and the presence of diverse xanthonoids and numerous gallotannins of a high polymerization degree, such as decamers (decagalloyl hexoses) and undecamers (undecagalloyl hexoses), which are reported for the first time for these *M. indica* cultivars, suggest the potential of these extracts for further studies. For instance, xanthonoids have been linked with anti-inflammatory and anticancer activities, and gallotannins of a higher degree of polymerization have been found to enhance such properties [22]; hence, structure–activity relationship studies would contribute to increase the knowledge on the fruits as a source of dietary compounds and bioactivities associated with potential health benefits.

Author Contributions: Conceptualization, M.N., E.A. and I.M.; Formal analysis, S.Q., G.A., K.W., and P.C.; Funding acquisition, M.N.; Investigation, M.N.; Writing—original draft, M.N., S.Q., G.A. and F.V.; Writing—review and editing, M.N., E.A., I.M., S.Q., G.A., K.W., F.V. and P.C.

Funding: This work was partially funded by a grant from the University of Costa Rica (115-B4-076).

Acknowledgments: Authors also thank the support from the Technological Institute of Costa Rica. Special thanks are due to Alonso Quesada from Costa Rican National Herbarium for his support with the vouchers.

References

1. Vithana, M.D.; Singh, Z.; Johnson, S.K. Regulation of the levels of health promoting compounds: Lupeol, mangiferin and phenolic acids in the pulp and peel of mango fruit: A review. *J. Sci. Food Agric.* **2019**, *99*, 3740–3751. [CrossRef] [PubMed]

2. Vega-Vega, V.; Silva-Espinoza, B.A.; Thalía Bernal-Mercado, A.; Vega-Vega, V.; Silva-Espinoza, B.A.; Cruz-Valenzuela, M.R.; Ayala-Zavala, J.F. Antimicrobial and antioxidant properties of byproduct extracts of mango fruit. *J. Appl. Bot. Food Qual.* **2013**, *86*, 205–211.

3. Carvalho, A.; Guedes, M.; De Souza, A.; Trevisan, M.; Lima, A.; Santos, F.; Rao, V. Gastroprotective Effect of Mangiferin, a Xanthonoid from Mangifera indica, against Gastric Injury Induced by Ethanol and Indomethacin in Rodents. *Planta Med.* **2007**, *73*, 1372–1376. [CrossRef] [PubMed]

4. Pourahmad, J.; Eskandari, M.R.; Shakibaei, R.; Kamalinejad, M. A Search for Hepatoprotective Activity of Fruit Extract of *Mangifera indica* L. against Oxidative Stress Cytotoxicity. *Plant Foods Hum. Nutr.* **2010**, *65*, 83–89. [CrossRef] [PubMed]

5. Stohs, S.J.; Swaroop, A.; Moriyama, H.; Bagchi, M.; Ahmad, T.; Bagchi, D. A Review on Antioxidant, Anti-Inflammatory and Gastroprotective Abilities of Mango (*Magnifera indica*) Leaf Extract and Mangiferin. *J. Nutr. Health Sci.* **2018**, *5*, 303. [CrossRef]

6. Luo, F.; Fu, Y.; Xiang, Y.; Yan, S.; Hu, G.; Huang, X.; Huang, G.; Sun, C.; Li, X.; Chen, K. Identification and quantification of gallotannins in mango (*Mangifera indica* L.) kernel and peel and their antiproliferative activities. *J. Funct. Foods* **2014**, *8*, 282–291. [CrossRef]

7. Abdullah, A.S.H.; Mohammed, A.S.; Rasedee, A.; Mirghani, M.E.S. Oxidative Stress-Mediated Apoptosis Induced by Ethanolic Mango Seed Extract in Cultured Estrogen Receptor Positive Breast Cancer MCF-7 Cells. *Int. J. Mol. Sci.* **2015**, *16*, 3528–3536. [CrossRef] [PubMed]

8. Noratto, G.D.; Bertoldi, M.C.; Krenek, K.; Talcott, S.T.; Stringheta, P.C.; Mertens-Talcott, S.U. Anticarcinogenic Effects of Polyphenolics from Mango (*Mangifera indica*) Varieties. *J. Agric. Food Chem.* **2010**, *58*, 4104–4112. [CrossRef] [PubMed]

9. Dai, J.; Mumper, R.J. Plant Phenolics: Extraction, Analysis and Their Antioxidant and Anticancer Properties. *Molecules* **2010**, *15*, 7313–7352. [CrossRef] [PubMed]

10. Quideau, S.; Deffieux, D.; Douat-Casassus, C.; Pouysegu, L.; Douat-Casassus, C. Plant Polyphenols: Chemical Properties, Biological Activities, and Synthesis. *Angew. Chem. Int. Ed.* **2011**, *50*, 586–621. [CrossRef] [PubMed]

11. Gold-Smith, F.; Fernandez, A.; Bishop, K. Mangiferin and Cancer: Mechanisms of Action. *Nutrients* **2016**, *8*, 396. [CrossRef] [PubMed]

12. Pan, L.L.; Wang, A.Y.; Huang, Y.Q.; Luo, Y.; Ling, M. Mangiferin induces apoptosis by regulating Bcl-2 and Bax expression in the CNE2 nasopharyngeal carcinoma cell line. *Asian Pac. J. Cancer Prev.* **2014**, *15*, 7065–7068. [CrossRef] [PubMed]

13. Fang, C.; Kim, H.; Noratto, G.; Sun, Y.; Talcott, S.T.; Mertens-Talcott, S.U. Gallotannin derivatives from mango (*Mangifera indica* L.) suppress adipogenesis and increase thermogenesis in 3T3-L1 adipocytes in part through the AMPK pathway. *J. Funct. Foods* **2018**, *46*, 101–109. [CrossRef]

14. García-Rivera, D.; Delgado, R.; Bougarne, N.; Haegeman, G.; Berghe, W.V. Gallic acid indanone and mangiferin xanthone are strong determinants of immunosuppressive anti-tumour effects of Mangifera indica L. bark in MDA-MB231 breast cancer cells. *Cancer Lett.* **2011**, *305*, 21–31. [CrossRef] [PubMed]

15. Berardini, N.; Reinhold, C.; Schieber, A. Characterization of Gallotannins and Benzophenone Derivatives from Mango (*Mangifera Indica* L.) Cv? Tommy Atkins Peels, Pulp and Kernels by High-Performance Liquid Chromatography/Electrospray Ionization Mass Spectrometry. *Rapid Commun. Mass Spectrom.* **2004**, *18*, 2208–2216. [CrossRef] [PubMed]

16. Schieber, A.; Berardini, N.; Carle, R. Identification of Flavonol and Xanthone Glycosides from Mango (*Mangifera indica* L. Cv. "Tommy Atkins") Peels by High-Performance Liquid Chromatography-Electrospray Ionization Mass Spectrometry. *J. Agric. Food Chem.* **2003**, *51*, 5006–5011. [CrossRef] [PubMed]

17. Berardini, N.; Fezer, R.; Conrad, J.; Beifuss, U.; Carle, R.; Schieber, A. Screening of Mango (*Mangifera indica* L.) Cultivars for Their Contents of Flavonol O–and Xanthone C-Glycosides, Anthocyanins, and Pectin. *J. Agric. Food Chem.* **2005**, *53*, 1563–1570. [CrossRef]

18. Arbizu-Berrocal, S.H.; Kim, H.; Fang, C.; Krenek, K.A.; Talcott, S.T.; Mertens-Talcott, S.U. Polyphenols from mango (*Mangifera indica* L.) modulate PI3K/AKT/mTOR-associated micro-RNAs and reduce inflammation in non-cancer and induce cell death in breast cancer cells. *J. Funct. Foods* **2019**, *55*, 9–16. [CrossRef]

19. Dorta, E.; González, M.; Lobo, M.G.; Sanchez-Moreno, C.; De Ancos, B. Screening of phenolic compounds in by-product extracts from mangoes (*Mangifera indica* L.) by HPLC-ESI-QTOF-MS and multivariate analysis for use as a food ingredient. *Food Res. Int.* **2014**, *57*, 51–60. [CrossRef]

20. Gómez-Caravaca, A.M.; López-Cobo, A.; Verardo, V.; Segura-Carretero, A.; Fernández-Gutiérrez, A. HPLC-DAD-Q-TOF-MS as a powerful platform for the determination of phenolic and other polar compounds in the edible part of mango and its by-products (peel, seed and seed husk). *Electrophoresi* **2016**, *37*, 1072–1084. [CrossRef]

21. Hu, K.; Dars, A.G.; Liu, Q.; Xie, B.; Sun, Z. Phytochemical profiling of the ripening of Chinese mango (*Mangifera indica* L.) cultivars by real-time monitoring using UPLC-ESI-QTOF-MS and its potential benefits as prebiotic ingredients. *Food Chem.* **2018**, *256*, 171–180. [CrossRef] [PubMed]

22. González-Sarrías, A.; Yuan, T.; Seeram, N.P. Cytotoxicity and structure activity relationship studies of maplexins A–I, gallotannins from red maple (*Acer rubrum*). *Food Chem. Toxicol.* **2012**, *50*, 1369–1376. [CrossRef] [PubMed]

23. Njoya, E.M.; Eloff, J.N.; McGaw, L.J. Croton gratissimus leaf extracts inhibit cancer cell growth by inducing caspase 3/7 activation with additional anti-inflammatory and antioxidant activities. *BMC Complement. Altern. Med.* **2018**, *18*, 305.

24. Prasetyaningrum, P.W.; Bahtiar, A.; Hayun, H. Synthesis and Cytotoxicity Evaluation of Novel Asymmetrical Mono-Carbonyl Analogs of Curcumin (AMACs) against Vero, HeLa, and MCF7 Cell Lines. *Sci. Pharm.* **2018**, *86*, 25. [CrossRef] [PubMed]

25. Prayong, P.; Barusrux, S.; Weerapreeyakul, N. Cytotoxic activity screening of some indigenous Thai plants. *Fitoterapia* **2008**, *79*, 598–601. [CrossRef] [PubMed]

26. Chothiphirat, A.; Nittayaboon, K.; Kanokwiroon, K.; Srisawat, T.; Navakanitworakul, R. Anticancer Potential of Fruit Extracts from Vatica diospyroides Symington Type SS and Their Effect on Program Cell Death of Cervical Cancer Cell Lines. *Sci. World J.* **2019**, *2019*, 1–9. [CrossRef] [PubMed]

27. Lavorgna, M.; Orlo, E.; Nugnes, R.; Piscitelli, C.; Russo, C.; Isidori, M. Capsaicin in Hot Chili Peppers: In Vitro Evaluation of Its Antiradical, Antiproliferative and Apoptotic Activities. *Plant Foods Hum. Nutr.* **2019**, *74*, 164–170. [CrossRef]

28. Navarro, M.; Zamora, W.; Quesada, S.; Azofeifa, G.; Alvarado, D.; Monagas, M. Fractioning of proanthocyanidins of Uncaria tomentosa. Composition and structure-bioactivity relationship. *Antioxidants* **2017**, *6*, 60. [CrossRef]

29. Singleton, V.; Rossi, J. Colorimetry of total phenolics with phosphomolybdic-phosphotungstic acid reagents. *Am. J. Enol. Vitic.* **1965**, *16*, 144–158.

30. Navarro, M.; Moreira, I.; Arnaez, E.; Quesada, S.; Azofeifa, G.; Vargas, F.; Alvarado, D.; Chen, P. Polyphenolic Characterization and Antioxidant Activity of Malus domestica and Prunus domestica Cultivars from Costa Rica. *Foods* **2018**, *7*, 15. [CrossRef]

31. Blois, M.S. Antioxidant Determinations by the Use of a Stable Free Radical. *Nature* **1958**, *181*, 1199–1200. [CrossRef]

32. Dávalos, A.; Gómez-Cordovés, C.; Bartolomé, B. Extending Applicability of the Oxygen Radical Absorbance Capacity (ORAC–Fluorescein) Assay. *J. Agric. Food Chem.* **2004**, *52*, 48–54. [CrossRef] [PubMed]

33. Gentile, C.; Di Gregorio, E.; Di Stefano, V.; Mannino, G.; Perrone, A.; Avellone, G.; Farina, V. Food quality and nutraceutical value of nine cultivars of mango (*Mangifera indica* L.) fruits grown in Mediterranean subtropical environment. *Food Chem.* **2019**, *277*, 471–479. [CrossRef] [PubMed]

34. Ramirez, J.; Zambrano, R.; Sepúlveda, B.; Simirgiotis, M. Antioxidant Properties and Hyphenated HPLC-PDA-MS Profiling of Chilean Pica Mango Fruits (*Mangifera indica* L. Cv. piqueño). *Molecules* **2013**, *19*, 438–458. [CrossRef] [PubMed]

35. Kim, H.; Moon, J.Y.; Kim, H.; Lee, D.S.; Cho, M.; Choi, H.K.; Kim, Y.S.; Mosaddik, A.; Cho, S.K. Antioxidant and antiproliferative activities of mango (*Mangifera indica* L.) flesh and peel. *Food Chem.* **2010**, *121*, 429–436. [CrossRef]

36. Monagas, M.; Garrido, I.; Lebrón-Aguilar, R.; Gómez-Cordovés, M.C.; Rybarczyk, A.; Amarowicz, R.; Bartolomé, B. Comparative Flavan-3-ol Profile and Antioxidant Capacity of Roasted Peanut, Hazelnut, and Almond Skins. *J. Agric. Food Chem.* **2009**, *57*, 10590–10599. [CrossRef] [PubMed]

37. Navarro, M.; Sanchez, F.; Murillo, R.; Martín, P.; Zamora, W.; Monagas, M.; Bartolomé, B. Phenolic assesment of *Uncaria tomentosa* L. (Cat's Claw): Leaves, stem, bark and wood extracts. *Molecules* **2015**, *20*, 22703–22717. [CrossRef]

38. Bartolomé, B.; Nuñez, V.; Monagas, M.; Gómez-Cordovés, C. In vitro antioxidant activity of red grape skins. *Eur. Food Res. Technol.* **2004**, *218*, 173. [CrossRef]

39. Navarro-Hoyos, M.; Alvarado-Corella, D.; Moreira-González, I.; Arnáez-Serrano, E.; Monagas-Juan, M. Polyphenolic Composition and Antioxidant Activity of Aqueous and Ethanolic Extracts from Uncaria tomentosa Bark and Leaves. *Antioxidants* **2018**, *7*, 65. [CrossRef]

40. Chen, J.F.; Song, Y.L.; Guo, X.Y.; Tu, P.F.; Jiang, Y. Characterization of the herb-derived components in rats following oral administration of Carthamus tinctokimrius extract by extracting diagnostic fragment ions (DFIs) in the MSn chromatograms. *Analyst* **2014**, *139*, 6474–6485. [CrossRef]

41. Morreel, K.; Saeys, Y.; Dima, O.; Lu, F.; Van De Peer, Y.; Vanholme, R.; Ralph, J.; Vanholme, B.; Boerjan, W. Systematic Structural Characterization of Metabolites in Arabidopsis via Candidate Substrate-Product Pair Networks. *Plant Cell* **2014**, *26*, 929–945. [CrossRef] [PubMed]

42. Callipo, L.; Cavaliere, C.; Fuscoletti, V.; Gubbiotti, R.; Samperi, R.; Laganà, A. Phenilpropanoate identification in young wheat plants by liquid chromatography/tandem mass spectrometry: Monomeric and dimeric compounds. *J. Mass Spectrom.* **2010**, *45*, 1026–1040. [CrossRef] [PubMed]

43. Barros, L.; Dueñas, M.; Pinela, J. Characterization and Quantification of Phenolic Compounds in Four Tomato (*Lycopersicon esculentum* L.) Farmers' Varieties in Northeastern Portugal Homegardens. *Plant Foods Hum. Nutr.* **2012**, *67*, 229. [CrossRef] [PubMed]

44. Zhang, H.; Cha, S.; Yeung, E.S. Colloidal Graphite-Assisted Laser Desorption/Ionization MS and MSnof Small Molecules. 2. Direct Profiling and MS Imaging of Small Metabolites from Fruits. *Anal. Chem.* **2007**, *79*, 6575–6584. [CrossRef] [PubMed]

45. Krenek, K.A.; Barnes, R.C.; Talcott, S.T. Phytochemical Composition and Effects of Commercial Enzymes on the Hydrolysis of Gallic Acid Glycosides in Mango (*Mangifera indica* L. cv. 'Keitt') Pulp. *J. Agric. Food Chem.* **2014**, *62*, 9515–9521. [CrossRef] [PubMed]

46. Li, X.; Duan, S.; Liu, H.; Xu, J.; Kong, M.; Liu, H.; Li, S. Sulfur-containing derivatives as characteristic chemical markers in control of sulfur-fumigated Moutan Cortex. *Yaoxue Xuebao* **2016**, *51*, 972–978. [CrossRef]

47. Clifford, M.N.; Stoupi, S.; Kuhnert, N. Profiling and Characterization by LC-MSnof the Galloylquinic Acids of Green Tea, Tara Tannin, and Tannic Acid. *J. Agric. Food Chem.* **2007**, *55*, 2797–2807. [CrossRef]

48. De Ancos, B.; Sánchez-Moreno, C.; Zacarías, L.; Rodrigo, M.J.; Sáyago Ayerdí, S.; Blancas Benítez, F.J.; Domínguez Avila, J.A.; Gonzalez-Aguilar, G.A. Effects of two different drying methods (freeze-drying and hot airdrying) on the phenolic and carotenoid profile of 'Ataulfo' mango by-products. *J. Food Meas. Charact.* **2018**, *12*, 2145–2157. [CrossRef]

49. Barreto, J.C.; Trevisan, M.T.S.; Hull, W.E.; Erben, G.; De Brito, E.S.; Pfundstein, B.; Würtele, G.; Spiegelhalder, B.; Owen, R.W. Characterization and Quantitation of Polyphenolic Compounds in Bark, Kernel, Leaves, and Peel of Mango (*Mangifera indica* L.). *J. Agric. Food Chem.* **2008**, *56*, 5599–5610. [CrossRef]

50. Beelders, T.; de Beer, D.; Stander, M.A.; Joubert, E. Comprehensive Phenolic Profiling of *Cyclopia genistoides* (L.) Vent. by LC-DAD-MS and -MS/MS Reveals Novel Xanthone and Benzophenone Constituents. *Molecules* **2014**, *19*, 11760–11790. [CrossRef]

51. Sobral, F.; Calhelha, R.C.; Barros, L.; Dueñas, M.; Tomás, A.; Santos-Buelga, C.; Vilas-Boas, M.; Ferreira, I.C.F.R. Flavonoid Composition and Antitumor Activity of Bee Bread Collected in Northeast Portugal. *Molecules* **2017**, *22*, 248. [CrossRef] [PubMed]

52. Fathoni, A.; Saepudin, E.; Cahyana, A.H.; Rahayu, D.U.C.; Haib, J. Identification of Nonvolatile Compounds in clove (*Syzygium aromaticum*) from Manado. In Proceedings of the 2nd International Symposium on Current Progress in Mathematics and Sciences 2016, Depok, Jawa Barat, Indonesia, 1–2 November 2016.

53. Justesen, U. Collision-induced fragmentation of deprotonated methoxylated flavonoids, obtained by electrospray ionization mass spectrometry. *J. Mass Spectrom.* **2001**, *36*, 169–178. [CrossRef] [PubMed]

54. López-Cobo, A.; Verardo, V.; Diaz-De-Cerio, E.; Segura-Carretero, A.; Fernández-Gutiérrez, A.; Gómez-Caravaca, A.M. Use of HPLC- and GC-QTOF to determine hydrophilic and lipophilic phenols in mango fruit (*Mangifera indica* L.) and its by-products. *Food Res. Int.* **2017**, *100*, 423–434. [CrossRef] [PubMed]

55. Ribeiro, S.; Barbosa, L.; Queiróz, J.; Knodler, M.; Schieber, A.; Barbosa, L.C.A. Phenolic compounds and antioxidant capacity of Brazilian mango (*Mangifera indica* L.) varieties. *Food Chem.* **2008**, *110*, 620–626. [CrossRef]

56. Shimamura, T.; Sumikura, Y.; Yamazaki, T.; Tada, A.; Kashiwagi, T.; Ishikawa, H.; Matsui, T.; Sugimoto, N.; Akiyama, H.; Ukeda, H. Applicability of the DPPH Assay for Evaluating the Antioxidant Capacity of Food Additives—Inter-laboratory Evaluation Study—. *Anal. Sci.* **2014**, *30*, 717–721. [CrossRef] [PubMed]

57. Turati, F.; Rossi, M.; Pelucchi, C.; Levi, F.; La Vecchia, C. Fruit and vegetables and cancer risk: A review of southern European studies. *Br. J. Nutr.* **2015**, *113*, S102–S110. [CrossRef]

58. La Vecchia, C.; Altieri, A.; Tavani, A. Vegetables, fruit, antioxidants and cancer: A review of Italian studies. *Eur. J. Nutr.* **2001**, *40*, 261–267. [CrossRef]

59. Abu Bakar, M.F.; Mohamed, M.; Rahmat, A.; Burr, S.A.; Fry, J.R. Cytotoxicity and polyphenol diversity in selected parts of Mangifera pajangand Artocarpus odoratissimusfruits. *Nutr. Food Sci.* **2010**, *40*, 29–38. [CrossRef]

60. Abbasi, A.M.; Liu, F.; Guo, X.; Fu, X.; Li, T.; Liu, R.H. Phytochemical composition, cellular antioxidant capacity and antiproliferative activity in mango (*Mangifera indica* L.) pulp and peel. *Int. J. Food Sci. Technol.* **2017**, *52*, 817–826. [CrossRef]

61. Abu Bakar, M.F.; Mohamad, M.; Rahmat, A.; Burr, S.A.; Fry, J.R. Cytotoxicity, cell cycle arrest, and apoptosis in breast cancer cell lines exposed to an extract of the seed kernel of Mangifera pajang (bambangan). *Food Chem. Toxicol.* **2010**, *48*, 1688–1697. [CrossRef]

62. Abdullah, A.S.H.; Mohammed, A.S.; Abdullah, R.; Mirghani, M.E.S.; Al-Qubaisi, M. Cytotoxic effects of *Mangifera indica* L. kernel extract on human breast cancer (MCF-7 and MDA-MB-231 cell lines) and bioactive constituents in the crude extract. *BMC Complement. Altern. Med.* **2014**, *14*, 199. [CrossRef] [PubMed]

63. Ganogpichayagrai, A.; Palanuvej, C.; Ruangrungsi, N. Antidiabetic and anticancer activities of Mangifera indica cv. Okrong leaves. *J. Adv. Pharm. Technol. Res.* **2017**, *8*, 19. [CrossRef] [PubMed]

64. Corrales-Bernal, A.; Urango, L.A.; Rojano, B.; Maldonado, M.E. In vitro and in vivo effects of mango pulp (*Mangifera indica* cv. Azucar) in colon carcinogenesis [Efectos in vitro e in vivo de la pulpa de mango (*Mangifera indica* cv. Azúcar) en la carcinogénesis de colon]. *Arch. Latinoam. Nutr.* **2014**, *64*, 16–23. [PubMed]

65. Adams, L.S.; Phung, S.; Yee, N.; Seeram, N.P.; Li, L.; Chen, S. Blueberry Phytochemicals Inhibit Growth and Metastatic Potential of MDA-MB-231 Breast Cancer Cells Through Modulation of the Phosphatidylinositol 3-Kinase Pathway. *Cancer Res.* **2010**, *70*, 3594–3605. [CrossRef] [PubMed]

66. Mahavorasirikul, W.; Viyanant, V.; Chaijaroenkul, W.; Itharat, A.; Na-Bangchang, K. Cytotoxic activity of Thai medicinal plants against human cholangiocarcinoma, laryngeal and hepatocarcinoma cells in vitro. *BMC Complement. Altern. Med.* **2010**, *10*, 55. [CrossRef] [PubMed]

67. Khurana, R.K.; Kaur, R.; Lohan, S.; Singh, K.K. Mangiferin: A promising anticancer bioactive. *Pharm. Pat. Anal.* **2016**, *5*, 169–181. [CrossRef]

68. Rajendran, P.; Rengarajan, T.; Nandakumar, N.; Divya, H.; Nishigaki, I. Mangiferin in cancer chemoprevention and treatment: Pharmacokinetics and molecular targets. *J. Recept. Signal Transduct.* **2015**, *35*, 76–84. [CrossRef] [PubMed]

69. Shoji, K.; Tsubaki, M.; Yamazoe, Y.; Satou, T.; Itoh, T.; Kidera, Y.; Tanimori, Y.; Yanae, M.; Matsuda, H.; Taga, A.; et al. Mangiferin induces apoptosis by suppressing Bcl-xL and XIAP expressions and nuclear entry of NF-κB in HL-60 cells. *Arch. Pharmacal Res.* **2011**, *34*, 469–475. [CrossRef]

70. Du Plessis-Stoman, D.; du Preez, J.G.H.; van de Venter, M. Combination treatment with oxaliplatin and mangiferin causes increased apoptosis and downregulation of NFKB in cancer cell lines. *Afr. J. Tradit. Complement. Altern. Med.* **2011**, *8*, 177–184. [CrossRef]

71. Li, H.; Huang, J.; Yang, B.; Xiang, T.; Yin, X.; Peng, W.; Cheng, W.; Wan, J.; Luo, F.; Li, H.; et al. Mangiferin exerts antitumor activity in breast cancer cells by regulating matrix metalloproteinases, epithelial to mesenchymal transition, and β-catenin signaling pathway. *Toxicol. Appl. Pharmacol.* **2013**, *272*, 180–190. [CrossRef]

72. Zhang, J.; Li, L.; Kim, S.-H.; Hagerman, A.E.; Lü, J. Anti-cancer, anti-diabetic and other pharmacologic and biological activities of penta-galloyl-glucose. *Pharm. Res.* **2009**, *26*, 2066–2080. [CrossRef] [PubMed]

73. Cai, Y.; Zhang, J.; Chen, N.G.; Shi, Z.; Qiu, J.; He, C.; Chen, M. Recent Advances in Anticancer Activities and Drug Delivery Systems of Tannins. *Med. Res. Rev.* **2017**, *37*, 665–701. [CrossRef] [PubMed]

74. Kawk, S.H.; Kang, Y.R.; Kim, Y.H. 1,2,3,4,6-Penta-O-galloyl-β-d-glucose suppresses colon cancer through induction of tumor suppressor. *Bioorg. Med. Chem. Lett.* **2018**, *28*, 2117–2123. [CrossRef] [PubMed]

75. Rothwell, J.A.; Knaze, V.; Zamora-Ros, R. Polyphenols: Dietary assessment and role in the prevention of cancers. *Curr. Opin. Clin. Nutr. Metab. Care* **2017**, *20*, 512–521. [CrossRef]

76. Zhou, Y.; Zheng, J.; Li, Y.; Xu, D.P.; Li, S.; Chen, Y.M.; Li, H.B. Natural Polyphenols for Prevention and Treatment of Cancer. *Nutrients* **2016**, *8*, 515. [CrossRef] [PubMed]

77. Lee, Y.J.; Jung, O.; Lee, J.; Son, J.; Cho, J.Y.; Ryou, C.; Lee, S.Y. Maclurin exerts anti-cancer effects on PC3 human prostate cancer cells via activation of p38 and inhibitions of JNK, FAK, AKT, and c-Myc signaling pathways. *Nutr. Res.* **2018**, *58*, 62–71. [CrossRef]

Article

A Comparative Study of Essential Oil Constituents and Phenolic Compounds of Arabian Lilac (*Vitex trifolia* var. Purpurea): An Evidence of Season Effects

Anahita Boveiri Dehsheikh [1], Mohammad Mahmoodi Sourestani [1,*], Paria Boveiri Dehsheikh [1], Sara Vitalini [2], Marcello Iriti [2,*] and Javad Mottaghipisheh [3]

[1] Department of Horticultural Science, Faculty of Agriculture, Shahid Chamran University of Ahvaz, Ahvaz 6135743311, Iran; anahitaboveiri84@gmail.com (A.B.D.); pboveiri17@gmail.com (P.B.D.)

[2] Department of Agricultural and Environmental Sciences, Milan State University, via G. Celoria 2, 20133 Milan, Italy; Sara.vitalini@unimi.it

[3] Department of Pharmacognosy, Faculty of Pharmacy, University of Szeged, 6720 Szeged, Hungary; Imanmottaghipisheh@pharmacognosy.hu

* Correspondence: mohammad.mahmoodi1360@gmail.com (M.M.S.); marcello.iriti@unimi.it (M.I.); Tel.: +98-6133364053 (M.M.S.); +39-0250316766 (M.I.)

Received: 8 January 2019; Accepted: 30 January 2019; Published: 2 February 2019

Abstract: To evaluate the fluctuation of secondary metabolites in Arabian lilac during a year, aerial parts of the plant were harvested in the middle of each month. The essential oils content from fresh and dried plant materials was analyzed by gas chromatography-flame ionization detector (GC-FID) and gas chromatography-mass spectrometer (GC-MS), individually. Phytochemical contents, along with antiradical scavenging potential of the related methanol extracts were separately assessed. The spring and autumn samples (fresh and dried) yielded more essential oil than the other samples. Forty-one compounds were identified totally in the oils and the major constituents characterized were β-caryophyllene, sabinene, and caryophyllene oxide. The extracts obtained from winter and summer plants possessed the highest total phenolics. The maximum amount of total flavonoid content was measured in winter (December and January), whereas the minimum one was observed in spring (March). The summer and winter samples showed the highest and lowest content of flavones and flavanols, respectively, whereas the anthocyanin content was higher in winter than in summer. Moreover, antiradical activity of the extracts in summer and winter samples was higher than in other seasons. Overall, this study can provide useful information regarding the best harvest period of Arabian lilac to yield the desired compounds for application in phytopharmaceutical and food industries.

Keywords: Verbenaceae; isoprenoids; β-caryophyllene; flavonoids; anthocyanins; antiradical capacity; DPPH

---

1. Introduction

The *Vitex* genus (Verbenaceae family) comprises nearly 270 species predominantly of trees and shrubs that are widely distributed in tropical and subtropical regions, along with some species growing in the temperate zones [1]. Many species of this genus have been extensively commended because of notable therapeutic effects on several female disorders such as endometriosis, abnormal menstrual cycle, menopausal conditions, corpus luteum insufficiency, hyper-prolactinaemia, infertility, acne, menopause, disrupted lactation, and cyclic breast pain, and, therefore, *Vitex* genus has been known as female herb since ancient times [2]. Furthermore, this genus possesses anti-inflammatory, anti-histaminic, anti-microbial, anti-pyretic, analgesic, and antioxidant activities and is used to remedy asthma, allergies, skin illnesses, diarrhea, and gastrointestinal and liver diseases. It was also reported these plants are insect repellent and aid in snake bite treatment [3]. The hydroalcoholic extract of *Vitex trifolia* exhibited the most potency against *Culex quinquefasciatus* larvae compared with other studied *Vitex* species [4].

The aforementioned benefits are attributed to the existence of a wide array of active substances including essential oils (EOs), phenolic acids, flavonoids, lignans, and anthraquinones, etc. [5].

Vitex trifolia L. is a stout aromatic shrub (less than 5 m height) with green-gray trifoliate leaves and white and violet flowers widespread in Southeast Asia, Micronesia, Australia, and East Africa [2,6]. *Vitex trifolia* var. purpurea commonly known as Arabian lilac is one of the most important varieties of this species. This plant can be easily distinguished from other plants of this species due to the annual color variation of leaves from green to violet and, therefore, it is mainly used as an ornamental plant in the landscapes. Leaves are traditionally recommended for the treatment of rheumatic pains, inflammation, sprains, and wound healing [1]. Infusion and decoction are also beneficial therapeutic preparations in the improvement of intestinal ailments and treatment of tuberculosis and amenorrhea and have been used by indigenous populations. Moreover, the essential oil (EO) of *Vitex trifolia* L. is often used as a sedative, an anti-inflammatory and used for headaches, colds and coughs as well as in liver disorders and HIV [1,7]. Phytochemical studies revealed that the methanolic extract of leaves possesses strong antioxidant activity and is considered as a potent anticancer agent due to its high content of phenolic compounds including phenolic acids, flavonoids, flavones, and flavonols [5].

In plants, the biosynthesis of secondary metabolites is strongly influenced by endogenous (genetic) and exogenous (environmental and edaphic) factors; therefore, their amount and chemical diversity are not constant during the plant lifecycle [8]. Environmental conditions, such as seasonal variations, are key factors capable of affecting the quantity and quality of these compounds.

There is a high correlation between the content and chemical composition of EOs and phenolic compounds and the antioxidant capacity of some medicinal plants with seasonal variations. For instance, the evaluation of the quantity and quality of EOs extracted from different species of *Ocimum* and *Thymus* genus, *Origanum syriacum*, *Mentha pulegium*, *Thymbra spicata*, *Satureja thymbra*, *Salvia trilobal*, and *Lanata camara* has shown a wide range of variations during the different months and seasons [9–15]. The same correlation was reported for total phenolic and flavonoid contents as well as the antioxidant capacity of the *Thymus* genus and *Vaccinium myrtillus* with the seasonal variations in environmental conditions [9,11,16]. However, to the best of our knowledge, to date, no information has been reported on the effect of seasonal variations on the quality and quality of secondary metabolites of *Vitex* genus. Hence, the present study focused, for the first time, on the variability of the chemical profile of Arabian lilac in different months of a year, to determine the best period of harvesting and achieve the highest level of desirable bioactive compounds for uses in pharmaceutical and food industries.

2. Materials and Methods

2.1. Plant Materials

Aerial parts of Arabian lilac (*Vitex trifolia* var. purpurea) were collected in the middle of each month during the year 2014 from Ahvaz, a city in the south of Khuzestan province, Iran (latitude 31° 20′ N, longitude 48° 40′ E, altitude 20 m asl) located in an arid to semi-arid region with temperate winter and very hot summer (Table 1). Plant material was taxonomically identified by the botanist Dr. Mehrangiz Chehrazi and voucher specimens were deposited in the herbarium (KHAU-235). One kilogram of plant material was collected from five trees close to each other during the year. Samples were harvested from all (four) sides of each plant. The samples were mixed; half (500 g) was used to extract EO and stored in liquid nitrogen to determine total anthocyanin content. The second portion (500 g) was shade dried at room temperature (20–25 °C) to extract the EO of dried aerial parts and measure total phenolic, flavonoid, flavone, and flavanol contents and antiradical capacity.

Table 1. The air temperature and relative humidity recorded in Iran, Ahvaz locality in 2014.

Month	Temperature (°C)			Relative Humidity (%)		
	Minimum	Maximum	Average	Minimum	Maximum	Average
January	4.6	21	12.8	29	92	61
February	9.6	23.2	16.4	19	64	42
March	15.4	30.4	22.9	27	94	61
April	15.2	28.2	21.7	20	46	33
May	28	37	32.5	19	43	31
June	28.8	45.2	37	8	31	20
July	30	50.2	40.1	12	62	37
August	30.6	48	39.3	20	74	47
September	29.6	44.8	37.2	15	46	31
October	24	35.6	29.8	41	97	69
November	10.6	28.4	19.5	25	73	49
December	14.4	17.6	16	64	95	80

2.2. Essential Oil Extraction

Fresh aerial parts (50 g) were individually subjected to hydro-distillation using Clevenger-type apparatus for 3 h according to the method recommended in the European Pharmacopoeia [17]. The obtained EOs were dried over anhydrous sodium sulfate (Na_2SO_4) and kept in sealed glass vials at 4 °C. The yields of EOs were determined based on fresh and dried matter and calculated as weight of oil (g) /100 g of fresh and dried aerial parts (% *w/w*), respectively.

2.3. Essential Oil Composition

The essential oils were analyzed by Agilent 7890 gas chromatograph coupled with an Agilent 5975 mass spectrometer; HP-5MS (5%-phenyl–95%-methyl polysiloxane) capillary column (30 m × 0.25 mm i.d., 0.25 μm film thickness, Agilent technologies, Santa Clara, CA, USA); helium carrier gas at 1.5 mL/min; injector temperature 280 °C; detector temperature 300 °C; column temperature 40 °C (1 min)–300 °C (3 min) at 5 °C/min. The injection was done with a split ratio of 10:1. Scanning (1 scan/s) was accomplished in the range 50 to 500 *m/z* using electron impact ionization at 70 eV. The gas chromatography-flame ionization detector (GC-FID) analyses were performed with a Varian 3800 gas chromatograph, equipped with a flame ionization detector and a capillary column CP-Sil8-CB (5%-diphenyl–95%-dimethyl polysiloxan, Agilent technologies, Santa Clara, CA, USA) of 30 m × 0.25 mm i.d., 0.25 μm film thickness, using the same conditions of the gas chromatography-mass spectrometer (GC-MS). The relative number of individual components of oil were calculated by the GC peak and arranged in order of GC elution. EOs constituents were identified based on the retention indices relative to C_5-C_{28} *n*-alkanes obtained in the same conditions and by comparing their mass spectra with those recorded in the Wiley 7 n.L and those reported in the literature [18].

2.4. Phenolic Compounds

2.4.1. Preparation of Plant Extracts

One gram of powdered dried leaves and 5 mL methanol (70%) were transferred to a centrifuge tube and shaken at 120 rpm for 24 h. Then, the samples were centrifuged at 4000 rpm for 15 min, and the supernatant was used for quantification of phenolic compounds. The obtained methanolic extract was diluted with deionized water at a ratio of 1 to 30 and stored at −20 °C.

2.4.2. Total Phenolic Content

Total phenolic content (TPC) was determined by the spectrophotometric method using the Folin-Ciocalteu reagent described by Wojdylo et al. [19]. Briefly, the methanolic extract (100 μL) was mixed with 200 μL of Folin-Ciocalteu's reagent (50%) and 2 mL of deionized water in a test tube. After 3 min, 1 mL of 20% Na_2CO_3 solution was added to the test tube and vortexed well. The mixture was maintained for 1 h at room temperature in the dark. The absorbance of the samples was recorded at 765 nm using a spectrophotometer (Shimadzu UV-1201, Kyoto, Japan). The concentration of total phenolics was calculated based on a standard curve of gallic acid (0, 50, 100, 150, 200, 250, 300, and 350 μg/mL) as a reference and expressed as mg GAE (gallic acid equivalents)/g of dry weight.

2.4.3. Total Flavonoid Content

The determination of total flavonoid content (TFC) was performed by the aluminum chloride colorimetric method [20]. For this purpose, 1 mL of the extracted sample solution was blended with 300 μL of $NaNO_2$ solution (5%). After 5 min, 600 μL of $AlCl_3$ (10%) was added to the reaction mixture that was allowed to remain for 6 min. Then, NaOH (4 mL, 1 M) was added to the sample solution and adjusted to 10 mL with distilled water. The absorbance was measured at 510 nm using a spectrophotometer. A calibration curve was constructed with quercetin solutions at concentrations 0 to 1000 μg/mL, and TFC was expressed in term of mg QUE (quercetin equivalents)/g of dry weight.

2.4.4. Total Flavone and Flavanol Contents

The total flavones and flavanols were assayed according to the modified method of Popova et al. [21]. In brief, the methanolic extract (1 mL) was mixed with 1 mL of $AlCl_3$ (5%), and the sample solution was adjusted to 2.5 mL with methanol (70%). The absorbance of samples was recorded at 425 nm using a spectrophotometer (Shimadzu UV-1201). Quercetin solution (0–35 μg/mL) was used as a reference standard, and total flavone and flavanol content was expressed as mg QUE/g of dry weight.

2.5. Total Anthocyanin Content

The assessment of total anthocyanin content (TAC) content was performed by the pH differential method [22]. Fresh leaves (3 g) were extracted with 20 mL solvent methanol: 10 N HCL (90:10%, *v/v*). The obtained extract was shaken at 200 rpm for 10 min, and the sample solution was centrifuged at 5000 rpm for 15 min at 4 °C. Then, 4 mL supernatant were separated and diluted with 36 mL of two different buffers; potassium chloride pH = 1.0 (0.025 M) and sodium acetate pH = 4.5 (0.4 M), respectively. After 20 min of incubation at room temperature and dark, the absorbance of samples was measured at 520 and 700 nm using a spectrophotometer (Shimadzu UV-1201). The concentration of anthocyanin calculated using Equations (1) and (2).

$$A = (A_{520} - A_{700})\, pH_{1.0} - (A_{520} - A_{700})\, pH_{4.5} \tag{1}$$

$$TAC = (A \times MW \times DF \times 100)/MA \tag{2}$$

where, A refers to the absorbance; MW is molecular weight of cyanidin-3-glucoside (C3G) (449.2 g/mol); DF is the dilution factor (10); MA is the molar absorptivity coefficient of cyanidin-3-glucoside (26,900 $M^{-1}cm^{-1}$), and TAC expressed as mg C3G/100 mL of plant extract.

2.6. Antioxidant Activity

The antioxidant activity of the extracts was estimated by the reduction of free radical DPPH (2,2-diphenyl-1-picrylhydrazyl) method, as described by Oke et al. [23] with minor modifications. The methanol extract (100 μL) was mixed with 5 mL of 0.004% methanol solution of DPPH radical. The reaction mixture was vigorously shaken, then incubated for 30 min in the dark at room temperature. The absorbance was read against a blank (methanol) at 517 nm using a Shimadzu UV-1201 spectrophotometer, and the percentage of antioxidant activity was calculated as Equation (3).

$$AA = [(A_B - A_S)/A_B] \times 100 \tag{3}$$

where, AA is antioxidant activity expressed as a percentage, A_B is absorbance of the blank (containing reaction mixture without sample extract) and A_S is absorbance of the reaction mixture with sample extract.

2.7. Statistical Analysis

All the experiments were performed in triplicate and data presented in this paper were statistically analyzed using completely randomized design (CRD) using SAS version 9.1 (SAS Institute Inc., Cary, NC, USA). When ANOVA F test was significant, Duncan's multiple range test was performed to determine the differences among the mean values at 5% level of significance and results are expressed as the mean $\pm$ standard deviation (SD).

3. Results

3.1. Essential Oil Content

The results demonstrated that the EO content underwent profound changes that seemed to strongly depend on the collection time ($p < 0.01$). EOs of the fresh aerial parts showed two intermittent ascendant trends during the year. The plants collected in spring and autumn contained a higher amount of EOs than the other seasons (Figure 1). The oil content of these samples began to increase from January and reached the highest amount (0.22%) at the beginning of the heat period (March). After that, the accumulation of EO gradually decreased from April, and the lowest amount was obtained in July (0.09%). There was a significant difference between the oil content of the plants collected from August to December. Similarly, the accumulation of oils increased again with a slight slope in August and September and showed the second peak in October (0.18%). Then, the percentage of EO declined from November and eventually reached its lowest level in December (0.10%). The dried aerial parts of plants contained more EOs than fresh ones varying in the range of 0.21 to 0.45%. Moreover, the essential oil content of the dried aerial parts significantly changed during the different seasons ($p < 0.01$) and showed a pattern similar to the one reported for the fresh aerial parts (Figure 1). The maximum amount of oils was detected in March (0.45%), followed by October and November (0.39%) while the minimum was recorded in August (0.21%).

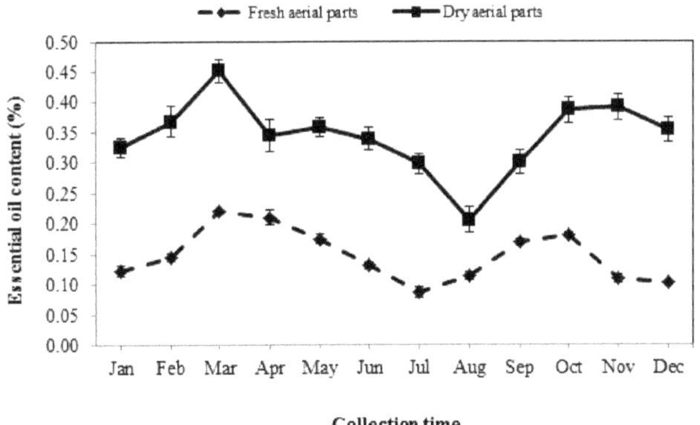

Figure 1. The seasonal variation of the essential oil content extracted from fresh and dried aerial parts of Arabian lilac. Values are the mean $\pm$ standard deviation (SD) of three replications ($n = 3$).

3.2. Essential Oil Composition

The EO composition of the fresh and dried aerial parts of Arabian lilac, along with their relative percentages and chemical classes in different collection periods are listed in Table 2. According to the GC-MS analysis, forty-one components were identified during different months, representing 97.02 to 98.92% and 97.03 to 99.96% of the fresh and dry oils, respectively. The major components of oils were classified into five groups; monoterpene hydrocarbons (α-pinene and sabinene), sesquiterpene hydrocarbons (β-caryophyllene and laurenene), oxygenated sesquiterpenes (caryophyllene oxide and (5E,9Z)-farnesyl-acetone), diterpene hydrocarbons (phytane and abietadiene) and oxygenated diterpenes (phytol, manool oxide, (6Z,10Z)-Pseudo phytol, manool, and 7α-hydroxy-manool).

The predominant compounds of the fresh and dried plant's oils were β-caryophyllene (22.60–35.03% and 25.14–32.43%, respectively), sabinene (7.24–16.73% and 11.04–18.38%, respectively), caryophyllene oxide (3.78–6.38% and 4.48–7.29%, respectively), (6Z,10Z)-Pseudo phytol (0.00–15.02% and 0.00–14.12%, respectively), laurenene (3.29–7.32% and 2.80–6.90%, respectively), (5E,9Z)-farnesyl-acetone (3.03–4.76% and 3.22–5.01%, respectively).

Sabinene, caryophyllene oxide, (6Z,10Z)-Pseudo phytol, (5E,9Z)-farnesyl-aceton, phytol, and α-pinene in fresh oils were less than in dried ones. In contrast, the fresh oils contained higher amounts of the β-caryophyllene, laurenene, phytane, manool oxide, and abietadiene compared with dried aerial parts. The most abundant chemical groups of the total fresh and dried sample's oils were sesquiterpenes, diterpenes, and monoterpenes. However, the fresh oils were richer in sesquiterpene and diterpene hydrocarbons, and the dry oils contained higher amounts of the oxygenated types of diterpenes and sesquiterpenes.

The data presented in Table 2 clearly shows a high variability in the percentages of the EO composition throughout the year. The biosynthesis of β-caryophyllene, as a major constituent of Arabian lilac's oil, markedly varied in different months and its annual fluctuation trend was the same for both fresh and dried oils. β-Caryophyllene in fresh oils had an ascendant trend from January, and the highest level was recorded in March (35.03%). Then, there was a clear decline during April and May. Moreover, the lowest level was obtained in June (22.60%). The amount of β-caryophyllene of dry oils increased continuously from February and declined again after rising in September. Furthermore, the maximum and minimum percentage of β-caryophyllene in dry oils were recorded in September (32.43%) and February (25.14%), respectively.

Sabinene, the second main component of Arabian lilac's oil, was highly increased in spring and autumn in fresh and dried material's oils. In contrary, there was a great decrease in its amount when oils were harvested during winter and summer. The highest content of sabinene in fresh and dried samples was found in March (16.73%) and October (18.38%), respectively. Additionally, the lowest amounts were obtained in July (7.24%) and August oils (11.04%), respectively (Table 2). The variation pattern of laurenene, manool oxide, manool, α-pinene, and abietadiene of fresh and dried aerial parts' oils was similar to that exhibited for β-caryophyllene and sabinene.

A quite different trend in the amount of caryophyllene oxide in fresh and dried sample's oils from other components was found. The amount of this constituent in the fresh plant's oils of summer and winter seasons was higher than in spring and autumn seasons and the maximum content was obtained in January (6.38%). The maximum of caryophyllene oxide of dried sample oils was observed in June (7.29%) whereas the least amount was obtained when plants were collected in September (4.48%). Similarly, the variations in the amount of (5E,9Z)-farnesyl-aceton, (6Z,10Z)-Pseudo phytol, phytol, and 7α-hydroxy-manool throughout the year was the same as those of caryophyllene oxide changes.

Table 2. chemical constituents of Arabian lilac (*Vitex trifolia* var. *purpurea*) essential oils at different collection periods (2014).

Compounds	RT[b]	KI[c]	KI[d]	Jan FAP	Jan DAP	Feb FAP	Feb DAP	Mar FAP	Mar DAP	Apr FAP	Apr DAP	May FAP	May DAP	Jun FAP	Jun DAP	Jul FAP	Jul DAP	Aug FAP	Aug DAP	Sep FAP	Sep DAP	Oct FAP	Oct DAP	Nov FAP	Nov DAP	Dec FAP	Dec DAP
				GC area (%)[a]																							
Thujene	7.58	933	924	0.15	0.37	0.18	0.38	0.23	0.36	0.29	0.36	0.48	0.60	0.23	0.51	0.17	0.34	0.16	0.27	0.21	0.20	0.26	0.42	0.14	0.35	0.11	0.20
α-Pinene	7.79	939	932	3.27	1.35	4.10	4.00	4.29	4.62	4.40	5.20	2.45	3.22	1.77	2.29	2.00	2.56	2.15	2.55	3.18	5.35	3.61	4.86	2.79	4.06	2.95	3.66
Sabinene	9.05	975	969	11.71	13.03	13.23	15.1	16.73	18.14	16.38	18.17	9.58	16.45	9.11	12.49	7.24	12.08	9.22	11.04	15.13	15.09	16.02	18.38	11.63	17.94	9.68	15.47
β-Pinene	9.09	976	974	1.54	0.78	2.00	0.80	1.86	1.07	0.47	1.14	0.50	0.96	0.37	0.07	0.31	0.04	0.32	0.03	0.57	0.01	0.87	0.20	1.83	0.86	1.26	0.30
1-Octen-3-ol	9.14	978	974	0.64	2.61	0.71	2.14	-	-	-	-	2.95	-	2.39	1.53	1.89	1.84	0.61	0.87	0.64	-	-	-	2.98	0.32	1.99	2.66
Myrcene	9.43	986	988	0.14	0.85	0.21	0.24	0.28	0.27	0.30	0.85	0.11	0.20	0.10	0.10	-	-	-	-	0.14	-	0.30	0.42	0.13	0.32	-	0.16
α-Terpinene	10.16	1009	1014	0.83	0.44	0.27	0.17	-	0.14	-	0.15	0.39	0.42	0.41	0.49	0.48	0.35	0.15	0.53	-	0.19	-	0.27	0.21	0.45	0.20	-
p-Cymene	10.39	1017	1020	-	0.47	0.24	0.75	0.42	0.78	0.55	0.53	0.28	0.78	0.24	0.37	0.16	0.27	-	-	0.21	-	-	1.07	0.15	0.51	0.10	-
β-Phellandrene	10.52	1021	1025	-	0.26	0.25	0.12	0.33	0.18	0.14	0.23	0.27	0.29	0.39	0.26	0.17	0.18	0.17	0.28	0.21	0.19	0.39	0.26	0.33	0.43	0.16	0.30
1,8-Cineole	10.60	1024	1026	0.10	0.19	0.22	-	0.16	-	-	0.12	0.25	0.21	0.21	0.23	0.24	0.18	0.14	0.31	0.16	0.14	0.27	0.23	0.26	0.19	-	0.22
γ-Terpinene	11.42	1052	1054	-	2.12	1.39	0.88	-	0.97	0.99	0.53	2.03	1.44	1.93	0.68	1.49	0.33	0.57	0.58	0.64	0.06	1.37	2.17	1.66	2.68	1.06	2.68
α-Terpinolene	12.28	1081	1086	-	0.22	-	0.26	-	0.12	-	0.09	0.10	0.10	0.08	-	-	-	-	-	-	-	0.11	0.18	0.35	0.24	-	-
Terpinen-4-ol	14.93	1171	1174	0.75	-	0.47	-	0.16	-	0.16	-	0.16	-	0.68	-	0.70	-	0.57	-	0.47	0.03	0.50	-	0.62	-	0.16	-
Estragol	15.48	1190	1195	-	2.72	-	1.68	0.63	-	-	-	2.88	-	1.33	-	1.76	-	-	0.98	1.68	-	-	-	0.98	-	0.27	2.87
Bornyl acetate	17.93	1280	1284	0.16	-	0.15	0.14	0.20	-	-	-	-	-	0.15	0.12	0.16	0.11	-	0.14	-	0.09	-	0.13	0.10	-	0.14	-
α-Terpinyl acetate	19.63	1343	1346	0.30	0.51	0.42	0.33	0.18	-	0.33	0.41	0.38	0.36	0.28	0.33	0.31	0.54	0.29	0.20	0.27	0.18	0.25	0.23	0.31	0.53	0.33	0.46
β-Caryophyllene	21.71	1422	1418	25.76	25.29	26.33	25.14	35.03	27.51	34.41	29.39	24.32	28.80	22.6	26.03	24.01	26.71	32.49	27.46	34.16	32.43	31.81	31.99	26.44	28.76	27.36	27.03
α-Caryophyllene	22.37	1450	1454	0.91	0.85	1.26	0.97	1.21	1.20	1.04	1.16	1.07	1.15	0.91	1.23	0.87	1.18	1.35	0.97	1.44	1.19	1.20	1.18	1.00	1.13	1.01	1.19
Germacrene-D	23.01	1476	1484	-	-	-	-	-	-	0.16	-	0.93	-	0.14	0.19	0.20	0.25	0.26	0.34	0.16	-	0.12	-	-	-	-	-
β-Selinene	23.14	1481	1489	-	0.10	-	-	-	-	-	-	0.78	-	0.53	0.21	0.23	0.29	0.42	0.38	0.25	-	-	-	-	-	0.17	-
α-Selinene	23.35	1490	1498	0.64	0.59	0.79	0.59	0.60	0.86	0.79	0.85	1.53	0.85	0.33	0.36	0.79	0.44	0.37	0.58	0.91	0.42	0.43	0.59	0.53	0.66	0.66	0.85
E-Nerolidol	24.91	1554	1561	1.18	-	-	0.11	-	0.16	-	-	0.33	0.15	0.22	0.17	0.19	0.19	0.23	0.13	0.22	0.10	-	-	0.14	-	0.11	-
Caryophyllene oxide	25.47	1578	1582	6.38	6.89	4.74	6.83	3.78	6.35	3.93	6.33	4.23	7.05	5.46	7.29	6.24	6.75	5.98	4.67	3.91	4.48	3.87	4.59	4.54	4.96	4.77	5.76
Caryophylladienol II	26.57	1626	1631	0.61	0.66	-	-	0.22	0.13	0.38	-	0.14	0.13	0.40	0.43	0.31	0.36	0.30	0.70	-	0.33	0.45	0.55	0.33	0.40	0.37	0.74
(Z)-14-hydroxy-Caryophyllene	27.41	1664	1666	0.91	1.82	0.71	0.63	0.42	-	0.58	-	1.81	0.75	2.46	1.34	2.75	1.25	2.63	0.87	1.85	0.62	0.45	0.52	0.39	0.72	0.55	0.71
Phytane	30.12	1789	1792	2.73	3.03	2.69	2.56	3.98	2.80	3.75	2.38	4.45	2.54	3.96	3.84	4.30	3.68	4.42	3.41	4.71	3.01	3.68	2.32	3.19	2.83	3.71	3.51
(5E,9Z)-Farnesyl acetone	31.93	1879	1886	4.76	5.01	4.49	4.10	3.12	3.22	3.41	3.29	3.23	3.37	4.25	4.11	4.61	4.42	4.49	4.88	3.55	4.28	3.03	4.11	4.09	3.49	4.49	4.04
Laurenene	32.26	1896	1887	4.07	2.80	3.71	3.92	5.26	4.42	5.46	4.56	3.29	4.10	3.56	3.57	4.06	3.67	6.88	5.89	7.32	6.90	6.23	6.89	3.68	5.77	3.90	4.00
epi-Laurenene	32.46	1906	1901	-	0.71	0.95	0.62	1.88	0.85	1.33	-	-	1.46	1.23	-	-	-	-	1.34	2.94	-	1.53	1.20	1.33	-	1.15	1.08
Phytol	33.26	1946	1942	4.01	3.43	3.72	4.58	3.01	3.81	2.28	3.31	2.83	3.55	2.99	3.77	3.81	4.51	3.71	4.03	2.08	3.34	2.51	3.02	2.55	3.97	3.90	3.17
Manool oxide	33.64	1965	1987	1.99	0.10	6.93	0.11	8.34	7.50	7.11	6.03	1.05	5.35	0.10	-	-	0.10	0.10	0.11	0.31	0.10	8.01	5.05	7.21	6.42	0.10	0.10
(6Z,10Z)-Pseudo phytol	34.03	1985	1988	10.01	0.23	0.20	0.64	-	-	0.10	6.72	0.11	8.34	10.92	12.67	15.02	14.12	9.93	13.54	0.10	13.63	0.27	0.34	0.40	0.28	10.5	0.15
Manoyl oxide	34.11	1989	1989	2.98	1.99	1.96	0.98	-	-	-	-	2.97	-	1.62	0.38	1.43	0.21	1.32	0.19	1.91	-	1.54	-	1.69	0.58	2.26	1.23
(E,E)-Geranyl linalool	34.88	2031	2026	0.11	1.96	1.97	1.91	-	1.02	-	-	2.99	-	1.92	0.17	1.49	0.19	0.23	0.13	0.22	-	-	-	0.14	-	1.21	1.09
Manool	35.37	2058	2056	1.35	0.74	1.45	1.24	1.74	1.45	1.67	1.30	1.50	1.32	1.37	0.60	-	0.90	0.10	0.89	1.36	1.09	1.30	1.26	1.44	1.19	1.62	1.12
Sclareolide	35.48	2064	2065	-	2.69	3.21	2.03	-	1.97	3.87	-	5.08	0.15	1.27	0.96	1.33	1.13	-	1.07	-	2.02	1.88	1.60	3.44	2.94	-	2.83
1-Octadecanol	35.64	2073	2077	0.56	2.86	0.84	1.98	-	-	-	-	2.99	-	2.95	1.43	2.73	1.92	2.29	1.89	0.55	-	-	0.88	1.99	0.49	0.93	2.48
Abietadiene	35.82	2082	2087	4.12	-	1.63	3.56	2.97	4.59	2.80	4.36	2.93	4.29	-	3.23	1.88	0.10	2.01	1.11	3.61	2.81	5.04	2.03	4.88	3.17	4.96	4.36
Laurenan-2-one	36.40	2114	2115	0.79	1.97	1.63	0.94	-	-	-	-	2.04	-	3.08	0.20	0.75	0.26	0.58	0.29	0.52	-	0.37	-	0.47	-	0.57	0.26
7α-hydroxy-Manool	38.62	2240	2237	1.58	3.06	0.53	1.73	-	0.31	0.11	1.58	2.25	1.58	2.92	4.18	1.63	2.80	0.93	2.48	0.36	1.66	0.31	1.97	0.35	1.28	1.42	2.07
Dehydroabietal	39.22	2276	2263	2.04	4.52	3.76	4.87	-	3.17	-	0.40	2.94	-	2.16	1.90	1.43	2.83	1.72	1.91	1.07	-	0.94	-	2.90	0.51	2.92	0.38
Chemical class																											
Monoterpene hydrocarbons				17.64	19.89	21.87	22.70	24.14	26.65	23.52	27.25	16.19	24.46	14.63	17.26	12.02	16.15	12.74	15.28	20.29	21.09	22.93	28.23	19.22	27.84	15.52	22.77
Oxygenated monoterpenes				1.31	0.70	1.26	0.47	0.70	-	0.49	0.53	0.79	0.57	1.32	0.68	1.41	0.83	1.00	0.65	0.90	0.44	1.02	0.59	1.29	0.72	0.63	0.68

Table 2. *Cont.*

Compounds	RT [b]	KI [c]	KI [d]	January		February		March		April		May		June		July		August		September		October		November		December	
				FAP	DAP	FAP	DAP	FAP	DAP	FAP	DAP	FAP	DAP	FAP	DAP	FAP	DAP	FAP	DAP	FAP	DAP	FAP	DAP	FAP	DAP	FAP	DAP
Total monoterpenes				18.95	20.59	23.13	23.17	24.84	26.65	24.01	27.78	16.98	25.03	15.95	17.94	13.43	16.98	13.74	15.93	21.19	21.53	23.95	28.82	20.51	28.56	16.15	23.45
Sesquiterpene hydrocarbons				31.38	30.34	33.04	31.24	43.98	34.84	43.19	35.96	31.92	36.36	29.30	31.59	30.16	32.54	41.77	36.96	47.18	40.94	41.32	41.85	32.98	36.32	34.25	34.15
Oxygenated sesquiterpenes				14.63	19.04	14.78	14.64	7.54	11.83	12.17	9.62	16.86	11.60	17.14	14.50	16.18	14.36	14.21	12.61	10.05	11.83	10.05	11.37	13.40	12.51	10.86	14.34
Total sesquiterpenes				46.01	49.38	47.82	45.88	51.52	46.67	55.36	45.58	48.78	47.96	46.44	46.09	46.34	46.90	55.98	49.57	57.23	52.77	51.37	53.22	46.38	48.83	45.11	48.4
Diterpene hydrocarbons				6.85	3.03	4.32	6.12	6.95	7.39	6.55	6.74	7.38	6.83	3.96	7.07	6.18	3.78	6.43	4.52	8.32	5.82	8.72	4.35	8.07	6.00	8.67	7.87
Oxygenated diterpenes				24.07	16.03	20.52	16.06	13.09	17.26	11.27	19.34	16.64	20.14	24.00	23.67	24.81	25.66	18.04	23.27	7.41	19.82	14.88	11.64	16.68	14.23	23.93	9.31
Total diterpenes				30.92	19.06	24.84	22.18	20.04	24.65	17.82	26.08	24.02	26.97	27.96	30.74	30.99	29.44	24.47	27.79	15.73	25.64	23.6	15.99	24.75	20.23	32.60	17.18
Phenylpropanoids				-	2.72	-	1.68	0.63	-	-	-	2.88	-	1.33	-	1.76	-	-	0.98	1.68	-	-	-	0.98	-	0.27	2.87
Alcohols				1.20	5.47	1.55	4.12	-	-	-	-	5.94	-	5.34	2.96	4.62	3.76	2.90	2.76	1.19	-	-	0.88	4.97	0.81	2.92	5.14
Total Identified				97.08	97.22	97.34	97.03	97.03	97.97	97.19	99.44	98.6	99.96	97.02	97.73	97.14	97.08	97.09	97.03	97.02	99.94	98.92	98.91	97.59	98.43	97.05	97.13
Not-identified				2.92	2.78	2.66	2.97	2.97	2.03	2.81	0.56	1.40	0.04	2.98	2.27	2.86	2.92	2.91	2.97	2.98	0.06	1.08	1.09	2.41	1.57	2.95	2.87

[a] Percentages obtained by FID peak-area normalization; RT [b] = Retention time; KI [c] = Kovats index; KI [d] = Kovats index in the literatures; FAP = Fresh aerial parts of plant; "-" = Not detected; DAP = Dried aerial parts of plant. The bold values are the main components of essential oil.

3.3. Total Phenolic Content

The total phenolic content (TPC) (*p* < 0.01) significantly varied in different months of the year ranging from 9.48 to 13.69 mg GAE/g of dry weight (DW) (Figure 2). The TPC gradually increased in December and reached its maximum amount in January (13.69 mg GAE/g DW). Then, a slight drop in TPC occurred in February (12.49 mg GAE/g DW) and March (11.19 mg GAE/g DW). Although, there was no significant difference among TPC of March, April, and May extracts, it increased slightly in spring, and the highest content was obtained in early summer (13.52 mg GAE/g DW). TPC dramatically declined from to September (9.48 mg GAE/g DW). The samples collected in autumn contained a lower value of TPC than those collected in other seasons. However, the least amount was recorded in September (9.48 mg GAE/g DW).

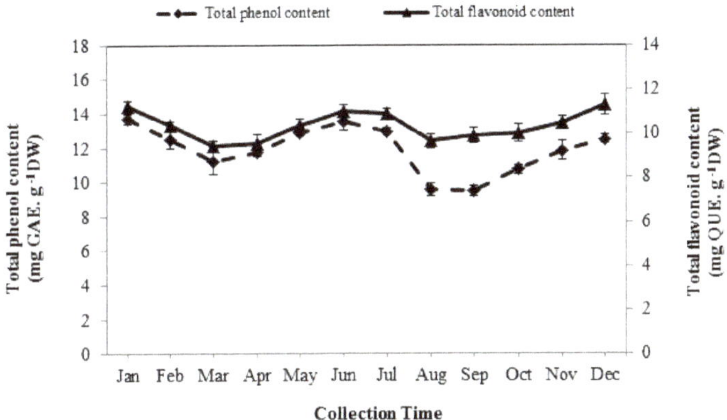

Figure 2. The seasonal variation of total phenol and flavonoid contents of Arabian lilac. Values are the mean ± SD of three replications (*n* = 3). DW: dry weight; GAE: gallic acid equivalents.

3.4. Total Flavonoid Content

The quantitative analysis of total flavonoid content (TFC) showed a significant difference in the diverse seasons (*p* < 0.01). As shown in Figure 2, there was a high level of TFC when the plants were harvested in the winter and summer, while its amount was low during the spring and autumn seasons. The maximum amount of TFC was found in December (11.31 mg QUE/g DW) and remained constant in January and, then, clearly decreased from the end of the winter season (February) and finally reached the minimum level in the middle of the spring (March) (9.40 mg QUE/g DW). Afterwards, a considerable rise occurred in June (11 mg QUE/g DW) which followed on July (10.92 mg QUE/g DW). However, this rate remained unchanged, and it began to increase again from September to December after a sharp drop in August (9.64 mg QUE/g DW).

3.5. Total Flavone and Flavanol Contents

The seasonal variation had marked impact on the amount of total flavone and flavanol contents (TFFC) (*p* < 0.01). The lowest concentration of TFFC was recorded in January (2.10 mg QUE/g DW). The amount of the TFFC was continuously increased from February, and the maximum concentration appeared in July (2.32 mg QUE/g DW). Thereafter, an opposite trend was observed and markedly declined from August to December (2.29–2.16 mg QUE/g DW) (Figure 3).

3.6. Total Anthocyanin Content

Total anthocyanin content (TAC) significantly differed (*p* < 0.01) from one month to another (1.26–2.81 mg C3G/100 mL). The pattern of seasonal variation of the TAC was interestingly quite reverse with respect to the TFFC pattern. The plants collected in winter contained higher levels of TAC compared with other seasons, and the maximum amount was observed in January (2.81 mg C3G/100 mL). Conversely, the anthocyanins were in lower concentrations during spring and especially summer. As shown in Figure 3, it declined from February to August with relatively sharp slope and the least level was obtained when the plants were harvested in August (1.26 mg C3G/100 mL), which was 2-fold less than that obtained in January. Then, the TAC increased during autumn from September to November.

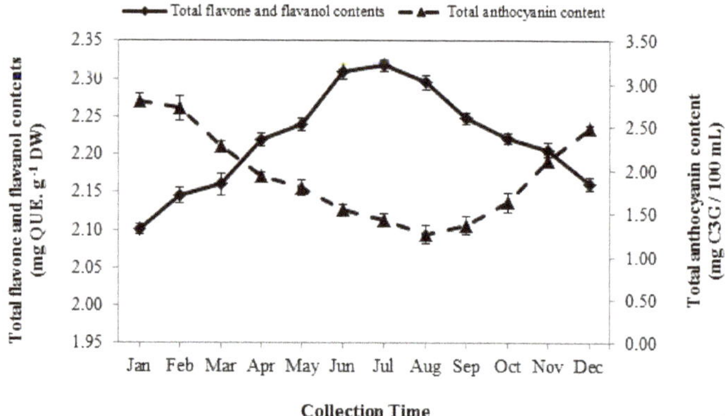

Figure 3. The seasonal variation of total flavone and flavanol, and anthocyanin content of Arabian lilac. Values are the mean ± SD of three replications (*n* = 3). QUE: quercetin equivalents.

3.7. Total Antioxidant Activity

Although the extracts wholly exhibited an effective reducing power of the radical species target DPPH•, their scavenging effect was strongly dependent on the time of collection ($p < 0.01$). The strongest antiradical potential was detected in January (91.02%) which was not constant, and it significantly decreased from February, and the extracts collected in March showed lower total antioxidant activity (TAA) (86.90%). The TAA of the extracts began to increase during spring and summer and peaked in July (90.15%). After that, a significant decrease trend was observed in the ability to scavenge free radicals of the extracts harvested in August and September. The TAA increased again from October to January (Figure 4). Eventually, the results obtained from the evaluation of DPPH• reducing power indicated that the extracts collected in winter and summer seasons were more effective than the ones harvested in spring and autumn.

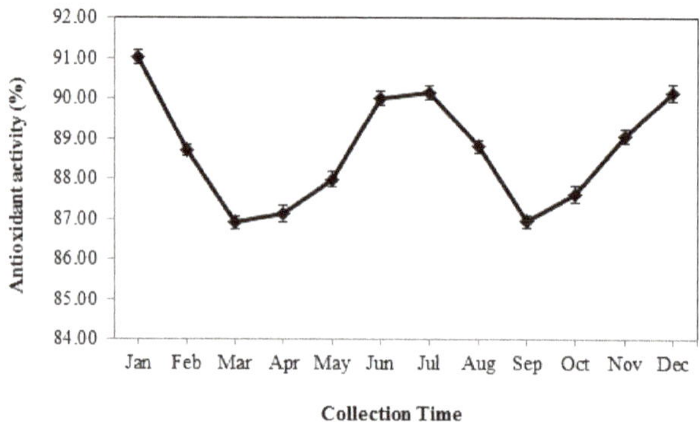

Figure 4. The seasonal variation of antioxidant activity of Arabian lilac. Values are the mean ± SD of three replications (*n* = 3).

4. Discussion

In the current study, the EO content extracted from fresh and dried aerial parts of Arabian lilac significantly varied in different collection periods and the samples harvested in spring and autumn showed higher EO contents than those collected during summer and winter. It seems that the increase of EO content (due to defense and protection roles) in the mentioned seasons might be related to the plant's exposure to stress because of a sudden change in environmental and climate conditions in the early spring (when the temperature begins to warm) and early autumn (when temperature begins to cool) (Figure 1). On the other hand, the decrease in EO content during the summer and winter seasons could be due to change in the biosynthetic pathway of secondary metabolites toward the production of metabolites with more defensive potential such as phenolic compounds. Several studies carried out on *Ocimum basilicum* [13], *Mentha spicata* and *Mentha pulegium* [24], and *Artemisia verlotiorum* [25] showed that environmental factors can considerably vary EO content of medicinal plants.

The variety of the EO's composition is influenced by some factors including radiation, humidity, soil condition, and temperature. Environmental stresses during seasonal variations can also alter the biosynthetic pathways [26]. In our findings, β-caryophyllene clearly decreased in summer and winter. Since β-caryophyllene is converted to its oxidized form, namely caryophyllene oxide, through an oxidation reaction, the reduction of the level of this component (in fresh and dried aerial parts' oils) and increase of caryophyllene oxide in these seasons could be ascribed to oxidative stress due to high and low temperature as well as the activation of enzymes of the oxidation reactions. Moreover, the opposite variation patterns of manool with other components of the same class and the reduction of their content in the fresh and dried plant's oils collected in summer and winter could be attributed to their conversion to oxidized and hydroxylated forms such as manool oxide and 7α-hydroxy-manool, respectively (Table 2). Previously published studies also revealed that EO composition of *Ocimum* species [10,13], *Rosemarinus officinalis* [27], and *Mentha spicata* [24] varies highly throughout the year.

Phenolic compounds, which have several hydroxyl groups, play a notable role in inactivating free radicals and exert antioxidant property [9]. The current study showed the total flavone, and flavanol contents of Arabian lilac significantly varied through the year and extract collected in summer, particularly on July, contained higher amounts of total flavones and flavanols (Figure 3). Regarding the ability of flavonoids to inhibit free radicals and the results obtained in this study, it seems that flavones and flavanols represent a defense mechanism of Arabian lilac against severe heat stress in the summer to reduce the harmful and destructive effects of the oxidative stress caused by heat stress. Our findings confirmed the results of Bujor et al. [16].

According to the literature, this is a first-time evaluation of changes in the TAC during the year. The TAC of the plant obtained in winter was significantly higher than in other seasons (Figure 3). A significant increase in the concentration of anthocyanins in the winter seems to be probably a defense mechanism against low temperature stress to reduce the damage of cell and tissue freezing [28–33]. The results interestingly revealed the variations pattern of anthocyanins was opposite with flavones and flavanols. Because the dihydro-flavonol is a common intermediate precursor of anthocyanins, flavones and flavanols [34], it seems that the decrease in temperature during winter has led to a change in the biosynthesis of flavones towards the production of anthocyanins, thereby reducing flavones and increasing anthocyanins.

Moreover, there were two peaks in the quantities of TPC and TFC during the summer and winter while their amount declined significantly in the spring and autumn (Figure 2). The increase in the TPC of extracts collected in summer and winter could be associated to the high quantity of flavones and flavanols content of Arabian lilac in summer and anthocyanins content in winter. Hence, the amount of TPC and TFC decreased in spring and autumn that could be related to a shift in the biosynthetic pathway of secondary metabolites towards the production of EOs in the plant. It seems that when the plant encounters oxidative stress due to a sudden change of temperature in the early spring and autumn, the biosynthetic pathway of secondary metabolites in the plant is directed toward the production of EOs rather than phenolic compounds [34,35].

Furthermore, there are few studies on seasonal variations of the antioxidant properties of medicinal plants. Galasso et al. [11] reported a significant change in the reducing power of *Thymus longicaulis* C. Presl extracts at various harvesting times. The seasonal variations of the antioxidant activity were similar to TPC and TFC (Figure 4). Indeed, the extracts yielded in summer and winter, especially in January and July, with higher levels of flavonoid compounds including flavones and flavanols, showed a higher antiradical power (Figure 2).

5. Conclusions

This study provides useful information on the seasonal variations of the quality and quantity of secondary metabolites as well as the antioxidant activity of the Arabian lilac. The results showed that weather conditions in the early spring and autumn were favorable to produce essential oil and increase the amounts of β-caryophyllene, sabinene, and manool. Although EO content decreased during the summer and winter seasons, the plants collected in these seasons contained higher caryophyllene oxide, phytol, (6Z,10Z)-Pseudo phytol, manool oxide and 7α-hydroxy-manool. The optimum time of harvesting plants to achieve the highest quantity of flavones and flavanols was summer, while the samples collected in the winter are rich in anthocyanins. It is recommended to harvest the plant in summer and winter, when it exhibits the highest TPC, TFC, and has the most powerful antioxidant activity. Therefore, the knowledge of the impact of environmental factors, such as seasonal changes, can help producers to choose the best period for plant harvesting and production of plant products richer in the desired compounds for exploitation in pharmaceutical and food industries.

Author Contributions: M.M.S. designed the experiments, harvested the materials and analyzed the essential oils. A.B.D. and P.B.D. performed the bioactivities. M.M.S. and A.B.D. wrote, and S.V., M.I., and J.M. revised the manuscript.

Funding: This project was financially supported by the Shahid Chamran University of Ahvaz.

Acknowledgments: The authors are thankful to Mehrangiz Chehrazi (Department of Horticultural Science, Faculty of Agriculture, Shahid Chamran University of Ahvaz, Ahvaz, Iran) for her excellent assist in taxonomical identification of plant material.

Conflicts of Interest: The authors declare no conflict of interest.

References

1. Radhapiyari Devi, W.; Singh, C.B. Chemical composition, anti-dermatophytic activity, antioxidant and total phenolic content within the leaves essential oil of *Vitex trifolia. Int. J. Phytocos. Nat. Ingred.* **2014**, *1*, 1–5. [CrossRef]
2. Rani, A.; Sharma, A. The genus *Vitex*: A review. *Pharmacogn. Rev.* **2013**, *7*, 188–198. [CrossRef] [PubMed]
3. Mohamed, M.A.; Abdou, A.M.; Hamed, M.M.; Saad, A.M. Characterization of bioactive phytochemical from the leaves of *Vitex trifolia. Int. J. Pharm. Appl.* **2012**, *3*, 419–428.
4. Kannathasan, K.; Senthilkumar, A.; Chandrasekaran, M.; Venkatesalu, V. Differential larvicidal efficacy of four species of *Vitex* against *Culex quinquefasciatus* larvae. *Parasitol. Res.* **2007**, *101*, 1721–1723. [CrossRef] [PubMed]
5. Yao, J.L.; Fang, S.M.; Liu, R.; Oppong, M.B.; Liu, E.W.; Fan, G.W.; Zhang, H. A Review on the Terpenes from Genus *Vitex. Molecules* **2016**, *21*, 1179. [CrossRef] [PubMed]
6. Tiwari, N.; Thakur, J.; Saikia, D.; Gupta, M.M. Antitubercular diterpenoids from *Vitex trifolia. Phytomedicine* **2013**, *20*, 605–610. [CrossRef] [PubMed]
7. Suksamrarn, A.; Werawattanametin, K.; Brophy, J.J. Variation of essential oil constituents in *Vitex trifolia* species. *Flavour Fragr. J.* **1991**, *6*, 97–99. [CrossRef]
8. Figueiredo, A.C.; Barroso, J.G.; Pedro, L.G.; Scheffer, J.J. Factors affecting secondary metabolite production in plants: Volatile components and essential oils. *Flavour Fragr. J.* **2008**, *23*, 213–226. [CrossRef]
9. Tohidi, B.; Rahimmalek, M.; Arzani, A. Essential oil composition, total phenolic, flavonoid contents, and antioxidant activity of *Thymus* species collected from different regions of Iran. *Food Chem.* **2017**, *220*, 153–161. [CrossRef]

10. Smitha, G.R.; Tripathy, V. Seasonal variation in the essential oils extracted from leaves and inflorescence of different *Ocimum* species grown in Western plains of India. *Ind. Crop. Prod.* **2016**, *94*, 52–64. [CrossRef]

11. Galasso, S.; Pacifico, S.; Kretschmer, N.; Pan, S.P.; Marciano, S.; Piccolella, S.; Bauer, R. Influence of seasonal variation on *Thymus longicaulis* C. Presl chemical composition and its antioxidant and anti-inflammatory properties. *Phytochemistry* **2014**, *107*, 80–90. [CrossRef] [PubMed]

12. Toncer, O.; Karaman, S.; Diraz, E. An annual variation in essential oil composition of *Origanum syriacum* from Southeast Anatolia of Turkey. *J. Med. Plant Res.* **2010**, *4*, 1059–1064. [CrossRef]

13. Hussain, A.I.; Anwar, F.; Sherazi, S.T.H.; Przybylski, R. Chemical composition, antioxidant and antimicrobial activities of basil (*Ocimum basilicum*) essential oils depends on seasonal variations. *Food Chem.* **2008**, *108*, 986–995. [CrossRef]

14. Randrianalijaona, J.A.; Ramanoelina, P.A.; Rasoarahona, J.R.; Gaydou, E.M. Seasonal and chemotype influences on the chemical composition of *Lantana camara* L.: Essential oil from Madagascar. *Anal. Chim. Acta* **2005**, *545*, 46–52. [CrossRef]

15. Muller-Riebau, F.J.; Berger, B.M.; Yegen, O.; Cakir, C. Seasonal variations in the chemical compositions of essential oils of selected aromatic plants growing wild in Turkey. *J. Agric. Food Chem.* **1997**, *45*, 4821–4825. [CrossRef]

16. Bujor, O.C.; Le Bourvellec, C.; Volf, I.; Popa, V.I.; Dufour, C. Seasonal variations of the phenolic constituents in bilberry (*Vaccinium myrtillus* L.) leaves, stems and fruits, and their antioxidant activity. *Food Chem.* **2016**, *213*, 58–68. [CrossRef] [PubMed]

17. Maisonneuve, S.A. *European Pharmacopoeia*, 3rd ed.; Council of Europe: Sainte-Ruffine, France, 1975.

18. Adams, R.P. Identification of essential oil components by gas chromatography/quadrupole mass spectrometry. *J. Am. Soc. Mass Spectrom.* **2007**, *16*, 1902–1903. [CrossRef]

19. Wojdylo, A.; Oszmianski, J.; Czemerys, R. Antioxidant activity and phenolic compound in 32 selected herbs. *Food Chem.* **2007**, *1005*, 940–949. [CrossRef]

20. Menichini, F.; Tundis, R.; Bonesi, M.; Loizzo, M.R.; Conforti, F.; Statti, G.; Di Cindi, B.; Houghton, P.J.; Menichini, F. The influence of fruit ripening on the phytochemical content and biological activity of *Capsicum chinense* Jacq. cv. Habanero. *Food Chem.* **2009**, *114*, 553–560. [CrossRef]

21. Popova, M.; Bankova, V.; Butovska, D.; Petkov, V.; Nikolova-Damyanova, B.; Sabatini, A.G.; Marcazzan, G.L.; Bogdanov, S. Validated methods for the quantification of biologically active constituents of poplar-type propolis. *Phytochem. Anal.* **2004**, *15*, 235–240. [CrossRef]

22. Lako, J.; Trenerry, V.C.; Wahlqvist, M.; Wattanapenpaiboon, N.; Sotheeswaran, S.; Premier, R. Phytochemical flavonols, carotenoids and the antioxidant properties of a wide selection of Fijian fruit, vegetables and other readily available foods. *Food Chem.* **2007**, *101*, 1727–1741. [CrossRef]

23. Oke, F.; Aslim, B.; Ozturk, S.; Altundag, S. Essential oil composition, antimicrobial and antioxidant activities of *Satureja cuneifolia* ten. *Food Chem.* **2009**, *112*, 874–879. [CrossRef]

24. Kofidis, G.; Bosabalidis, A.; Kokkini, S. Seasonal variations of essential oils in a linalool-rich chemotype of *Mentha spicata* grown wild in Greece. *J. Essent. Oil Res.* **2006**, *16*, 469–472. [CrossRef]

25. Chericoni, S.; Flamini, G.; Campeol, E.; Cioni, P.L.; Morelli, I. GC-MS analysis of the essential oil from the aerial parts of *Artemisia verlotiorum*: Variability during the year. *Biochem. Syst. Ecol.* **2004**, *32*, 423–429. [CrossRef]

26. Pazouki, L.; Kanagendran, A.; Li, S.; Kannaste, A.; Memari, H.R.; Bichele, R.; Niinemets, U. Mono-and sesquiterpene release from tomato (*Solanum lycopersicum*) leaves upon mild and severe heat stress and through recovery: From gene expression to emission responses. *Environ. Exp. Bot.* **2016**, *132*, 1–5. [CrossRef] [PubMed]

27. Celiktas, O.Y.; Kocabas, E.E.H.; Bedir, E.; Sukan, F.V.; Ozek, T.; Baser, K.H.C. Antimicrobial activities of methanol extracts and essential oils of *Rosmarinus oficinalis*, depending on location and seasonal variations. *Food Chem.* **2007**, *100*, 553–559. [CrossRef]

28. Mori, K.; Goto-Yamamoto, N.; Kitayama, M.; Hashizume, K. Loss of anthocyanins in red-wine grape under high temperature. *J. Exp. Bot.* **2007**, *58*, 1935–1945. [CrossRef]

29. Ubi, B.W.; Honda, C.; Bessho, H.; Kondo, S.; Wada, M.; Kobayashi, S.; Moriguchi, T. Expression analysis of anthocyanin biosynthetic genes in apple skin: Effect of UV-B and temperature. *Plant Sci.* **2006**, *170*, 571–578. [CrossRef]

30. Lo Piero, A.R.; Puglisi, I.; Rapisarda, P.; Petrone, G. Anthocyanins accumulation and related gene expression in red orange fruit induced by low temperature storage. *J. Agric. Food Chem.* **2005**, *53*, 9083–9088. [CrossRef]

31. Dela, G.; Or, E.; Ovadia, R.; Nissim-Levi, A.; Weiss, D.; Oren-Shamir, M. Changes in anthocyanin concentration and composition in 'Jaguar' rose flowers due to transient high temperature conditions. *Plant Sci.* **2003**, *164*, 333–340. [CrossRef]

32. Shvarts, M.; Borochov, A.; Weiss, D. Low temperature enhances petunia flower pigmentation and induces chalcone synthase gene expression. *Physiol. Plant* **1997**, *99*, 67–72. [CrossRef]

33. Leyva, A.; Jarillo, J.; Salinas, J.; Martinez-Zapater, M. Low temperature induces the accumulation of phenylalanine ammonialyase and chalcone synthase mRNAs of *Arabidopsis thaliana* in a light-dependent manner. *Plant Physiol.* **1995**, *108*, 39–46. [CrossRef] [PubMed]

34. Gutzeit, H.O.; Ludwig-Muller, J. *Plant Natural Products: Synthesis, Biological Functions and Practical Applications*; John Wiley and Sons: New York, NY, USA, 2014; pp. 19–21.

35. Borrelli, G.M.; Trono, D. Molecular Approaches to genetically improve the accumulation of health-promoting secondary metabolites in staple crops-A case study: The lipoxygenase-B1 genes and regulation of the carotenoid content in pasta products. *Int. J. Mol. Sci.* **2016**, *17*, 1177. [CrossRef] [PubMed]

MDPI

St. Alban-Anlage 66

4052 Basel

Switzerland

Tel. +41 61 683 77 34

Fax +41 61 302 89 18

www.mdpi.com

Foods Editorial Office

E-mail: foods@mdpi.com

www.mdpi.com/journal/foods